T0189233

Advances in
High Performance Computing
and Computational Sciences

The 1st Kazakh-German
Advanced Research Workshop, Almaty,
Kazakhstan, September 25 to October 1, 2005

Yurii Shokin
Michael Resch
Nargozy Danaev
Murat Orunkhanov
Nina Shokina
(Editors)

 Springer

Prof. Dr. Yurii Shokin
Institute of Computational Technologies of SB RAS
Ac. Lavrentyev Ave. 6
630090 Novosibirsk
Russia

Prof. Dr. Michael Resch
Dr. Nina Shokina
High Performance Computing Center Stuttgart (HLRS)
University of Stuttgart
Nobelstraße 19
70569 Stuttgart
Germany

Prof. Dr. Nargozy Danaev
Prof. Dr. Murat Orunkhanov
Institute of Mathematics and Mechanics
Al-Farabi Kazakh National University
Masanchi str. 39/47
050012 Almaty
Kazakhstan

ISBN 978-3-642-07039-6   e-ISBN 978-3-540-33844-4

Springer is a part of Springer Science+Business Media
springer.com

© Springer-Verlag Berlin Heidelberg 2006
Softcover reprint of the hardcover 1st edition 2006

NNFM Editor Addresses

Prof. Dr. Ernst Heinrich Hirschel
(General editor)
Herzog-Heinrich-Weg 6
D-85604 Zorneding
Germany
E-mail: e.h.hirschel@t-online.de

Prof. Dr. Kozo Fujii
Space Transportation Research Division
The Institute of Space
and Astronautical Science
3-1-1, Yoshinodai, Sagamihara
Kanagawa, 229-8510
Japan
E-mail: fujii@flab.eng.isas.jaxa.jp

Dr. Werner Haase
Höhenkirchener Str. 19d
D-85662 Hohenbrunn
Germany
E-mail: werner@haa.se

Prof. Dr. Bram van Leer
Department of Aerospace Engineering
The University of Michigan
Ann Arbor, MI 48109-2140
USA
E-mail: bram@engin.umich.edu

Prof. Dr. Michael A. Leschziner
Imperial College of Science
Technology and Medicine
Aeronautics Department
Prince Consort Road
London SW7 2BY
U. K.
E-mail: mike.leschziner@ic.ac.uk

Prof. Dr. Maurizio Pandolfi
Politecnico di Torino
Dipartimento di Ingegneria
Aeronautica e Spaziale
Corso Duca degli Abruzzi, 24
I-10129 Torino
Italy
E-mail: pandolfi@polito.it

Prof. Dr. Jacques Periaux
Dassault Aviation
78, Quai Marcel Dassault
F-92552 St. Cloud Cedex
France
E-mail: jperiaux@free.fr

Prof. Dr. Arthur Rizzi
Department of Aeronautics
KTH Royal Institute of Technology
Teknikringen 8
S-10044 Stockholm
Sweden
E-mail: rizzi@kth.se

Dr. Bernard Roux
L3M – IMT La Jetée
Technopole de Chateau-Gombert
F-13451 Marseille Cedex 20
France
E-mail: broux@l3m.univ-mrs.fr

Prof. Dr. Yurii I. Shokin
Siberian Branch of the
Russian Academy of Sciences
Institute of Computational
Technologies
Ac. Lavrentyeva Ave. 6
630090 Novosibirsk
Russia
E-mail: shokin@ict.nsc.ru

Preface

This volume is published as the proceedings of the first Kazakh-German Advanced Research Workshop on Computational Science and High Performance Computing in Almaty, Kazakhstan, on September 25 - October 1, 2005.

The contributions of these proceedings were provided and edited by the authors, chosen after a careful selection and reviewing process.

The workshop was organized by the High Performance Computing Center Stuttgart (Stuttgart, Germany), al-Farabi Kazakh National University (Almaty, Kazakhstan) and the Institute of Computational Technologies SB RAS (Novosibirsk, Russia) in the framework of activities of the German-Russian Center for Computational Technologies and High Performance Computing.

In March 2005, further to the long-term collaboration between German and Siberian scientists, at Prof. Yurii Shokin's suggestion the Kazakh scientists from the al-Farabi Kazakh National University and Institute of Mathematics and Mechanics (al-Farabi Kazakh National University) have participated in the second Russian-German Advanced Research Workshop on Computational Science and High Performance Computing in Stuttgart, hereby establishing a multilateral cooperation. A keen interest has been shown in developing a close cooperation between German and Kazakh specialists in the field of computational science and high performance computing, giving the possibility of sharing and discussing the latest results and developing further scientific contacts.

The topics of the workshop include numerical modelling in problems on non-linear fiber optics and problems of electrical sounding, high performance computing, numerical modelling of flows in hydro turbines, computational fluid dynamics, visualization of computational modelling results, theory of mathematical methods, numerical modelling in problems on flame propagation, spray combustion and supersonic turbulent jets, smoothed particle hydrodynamics, numerical modelling in industrial problems, large-eddy simulations (LES) of complex flows.

The participation of representatives of major research organizations engaged in the solution of the most complex problems of mathematical modelling, development of new algorithms, programs and key elements of information technologies, elaboration and implementation of software and hardware for high performance computing systems, provided a high level of competence of the workshop.

Among the Kazakh participants were researchers of the al-Farabi Kazakh National University (Almaty), the Institute of Mathematics and Mechanics (al-Farabi Kazakh National University, Almaty), the Institute of Mathematics (Almaty).

Among the German participants were the heads and leading specialists of the High Performance Computing Center Stuttgart (HLRS) (University of Stuttgart), the Institute of Aerodynamics and Gasdynamics (University of Stuttgart), the Institute of Aerodynamics (RWTH Aachen), the Institute of Applied Mathematics (University of Freiburg i. Br.), the Institute of Technical Thermodynamics (University of Karlsruhe(TH)).

Among the Russian participants were researchers of the Institute of Computational Technologies SB RAS (Novosibirsk) and the Sobolev Institute of Mathematics SB RAS (Novosibirsk).

This volume provides state-of-the-art scientific papers, presenting the latest results of the leading German, Kazakh and Russian institutions.

We are glad to see the successful continuation and promising perspectives of the highly professional international scientific meetings, which bring new insights and show the ways of future development in the problems of computational sciences and information technologies.

The editors would like to express their gratitude to all the participants of the workshop and wish them a further successful and fruitful work.

Novosibirsk and Stuttgart,   *Yurii Shokin*
November 2005   *Michael Resch*
                                                         Nargozy Danaev
                                                      Murat Orunkhanov
                                                          Nina Shokina

List of Contributors

N.A. Alkishriwi
Institute of Aerodynamics
RWTH Aachen
Wuelnnerstr. zw. 5 u.7
Aachen, 52062, Germany
office@aia.rwth-aachen.de

G. Balakayeva
Institute of Mathematics and
Mechanics
al-Farabi Kazakh National University
Masanchi str. 39/47
Almaty, 480012, Kazakhstan
balakayeva@kazsu.kz

Y. Bogdanov
Institute of Mathematics and
Mechanics
al-Farabi Kazakh National University
Masanchi str. 39/47
Almaty, 480012, Kazakhstan
j@dasm.kz

T. Bönisch
High Performance Computing Center
Stuttgart (HLRS)
University of Stuttgart
Nobelstraße 19
Stuttgart, 70569, Germany
boenish@hlrs.de

A. Burri
Institute of Applied Mathematics
University of Freiburg i. Br.
Hermann-Herder-Str. 10
Freiburg i. Br., 79104, Germany
burriad@mathematik.uni-
freiburg.de

S.G. Cherny
Institute of Computational Tech-
nologies SB RAS
Lavrentiev Ave. 6
Novosibirsk, 630090, Russia
cher@ict.nsc.ru

D.V. Chirkov
Institute of Computational Tech-
nologies SB RAS
Lavrentiev Ave. 6
Novosibirsk, 630090, Russia
dchirkov@ngs.ru

G. Dairbaeva
al-Farabi Kazakh National University
Al-Farabi av. 71
Almaty, 050078, Kazakhstan
dairbaeva@kazsu.kz

A. Dedner
Institute of Applied Mathematics
University of Freiburg i. Br.

Hermann-Herder-Str. 10
Freiburg i. Br., 79104, Germany
dedner@mathematik.uni-
freiburg.de

D. Diehl
Institute of Applied Mathematics
University of Freiburg i. Br.
Hermann-Herder-Str. 10
Freiburg i. Br., 79104, Germany
dennis@mathematik.uni-
freiburg.de

M.P. Fedoruk
Institute of Computational Tech-
nologies SB RAS
Lavrentiev Ave. 6
Novosibirsk, 630090, Russia
mife@ict.nsc.ru

S. Ganzenmüller
Department of Computer Engineer-
ing
University of Tübingen
Sand 13
Tübingen, 72076, Germany
ganzenmu@informtik.uni-
tuebingen.de

M. Hipp
Department of Computer Engineer-
ing
University of Tübingen
Sand 13
Tübingen, 72076, Germany
hippm@informtik.uni-tuebingen.de

S. Holtwick
Institute of Theoretical Astrophysics
University of Tübingen
Auf der Morgenstelle 10
Tübingen, 72076, Germany
holtwick@tat.physik.uni-
tuebingen.de

S.I. Kabanikhin
Sobolev Institute of Mathematics SB
RAS
Lavrentiev Ave. 4
Novosibirsk, 630090, Russia
Kazakh–British Technical University
Tolebi ave. 59
Almaty, 050091, Kazakhstan
kabanikh@math.nsc.ru

A. Kaltayev
al-Farabi Kazakh National University
Al-Farabi av. 71
Almaty, 050078, Kazakhstan
kaltayev@kazsu.kz

G.S. Khakimzyanov
Institute of Computational Tech-
nologies SB RAS
Lavrentiev Ave. 6
Novosibirsk, 630090, Russia
khak@ict.nsc.ru

R. Klöfkorn
Institute of Applied Mathematics
University of Freiburg i. Br.
Hermann-Herder-Str. 10
Freiburg i. Br., 79104, Germany
robertk@mathematik.uni-
freiburg.de

U. Küster
High Performance Computing Center
Stuttgart (HLRS)
University of Stuttgart
Nobelstraße 19
Stuttgart, 70569, Germany
kuester@hlrs.de

V.N. Lapin
Institute of Computational Tech-
nologies SB RAS
Lavrentiev Ave. 6
Novosibirsk, 630090, Russia
lapvas@ngs.ru

A.S. Lebedev
Institute of Computational Technologies SB RAS
Lavrentiev Ave. 6
Novosibirsk, 630090, Russia
sasa@ict.nsc.ru

S. Lipp
Institute of Technical Thermodynamics
University of Karlsruhe(TH)
Kaiserstraße 12
76131 Karlsruhe, Germany
lipp@itt.mach.uni-karlsruhe.de

I.F. Lobareva
Institute of Computational Technologies SB RAS
Lavrentiev Ave. 6
Novosibirsk, 630090, Russia
irenek@ngs.ru

U. Maas
Institute of Technical Thermodynamics
University of Karlsruhe(TH)
Kaiserstraße 12
76131 Karlsruhe, Germany
maas@itt.mach.uni-karlsruhe.de

M. Meinke
Institute of Aerodynamics
RWTH Aachen
Wuelnnerstr. zw. 5 u.7
Aachen, 52062, Germany
office@aia.rwth-aachen.de

B. Mukanova
Institute of Mathematics and Mechanics
al-Farabi Kazakh National University
Masanchi str. 39/47
Almaty, 480012, Kazakhstan
mubaga@kazsu.kz

A.Zh. Naimanova
Institute of Mathematics
Pushkin str. 125
Almaty, 480100, Kazakhstan
ked@math.kz

M. Ohlberger
Institute of Applied Mathematics
University of Freiburg i. Br.
Hermann-Herder-Str. 10
Freiburg i. Br., 79104, Germany
mario@mathematik.uni-freiburg.de

M. Orunkhanov
Institute of Mathematics and Mechanics
al-Farabi Kazakh National University
Masanchi str. 39/47
Almaty, 480012, Kazakhstan
mubaga@kazsu.kz

S. Pinkenburg
Department of Computer Engineering
University of Tübingen
Sand 13
Tübingen, 72076, Germany
pinkenbu@informtik.uni-tuebingen.de

K.B. Rakhmetova
al-Farabi Kazakh National University
Al-Farabi av. 71
Almaty, 050078, Kazakhstan
nich7@kazsu.kz

G.I. Ramazanova
al-Farabi Kazakh National University
Al-Farabi av. 71
Almaty, 050078, Kazakhstan
nich7@kazsu.kz

M. Resch
High Performance Computing Center
Stuttgart (HLRS)
University of Stuttgart
Nobelstraße 19
Stuttgart, 70569, Germany
resch@hlrs.de

W. Rosenstiel
Department of Computer Engineering
University of Tübingen
Sand 13
Tübingen, 72076, Germany
rosen@informtik.uni-tuebingen.de

H. Ruder
Institute of Theoretical Astrophysics
University of Tübingen
Auf der Morgenstelle 10
Tübingen, 72076, Germany
ruder@tat.physik.uni-tuebingen.de

A. Ruprecht
Institute for Fluid Mechanics and
Hydraulic Machinery
University of Stuttgart
Pfaffenwaldring 10
Stuttgart, 70550, Germany
ruprecht@ihs.uni-stuttgart.de

W. Schröder
Institute of Aerodynamics
RWTH Aachen
Wuelnnerstr. zw. 5 u.7
Aachen, 52062, Germany
office@aia.rwth-aachen.de

S.V. Sharov
Institute of Computational Technologies SB RAS
Lavrentiev Ave. 6
Novosibirsk, 630090, Russia
serge@ict.nsc.ru

K. Shakenov
al-Farabi Kazakh National University
Al-Farabi av. 71
Almaty, 050078, Kazakhstan
shakenov2000@mail.ru

Yu.I. Shokin
Institute of Computational Technologies SB RAS
Lavrentiev Ave. 6
Novosibirsk, 630090, Russia
shokin@ict.nsc.ru

N.Yu. Shokina
High Performance Computing Center
Stuttgart (HLRS)
University of Stuttgart
Nobelstraße 19
Stuttgart, 70569, Germany
shokina@hlrs.de

O.V. Shtyrina
Institute of Computational Technologies SB RAS
Lavrentiev Ave. 6
Novosibirsk, 630090, Russia
shtyrinaov@ngs.ru

V.A. Skorospelov
Sobolev Institute of Mathematics SB RAS
Lavrentiev Ave. 4
Novosibirsk, 630090, Russia
vskrsp@math.nsc.ru

R. Stauch
Institute of Technical Thermodynamics
University of Karlsruhe(TH)
Kaiserstraße 12
76131 Karlsruhe, Germany
stauch@itt.mach.uni-karlsruhe.de

Zh. Ualiev
al-Farabi Kazakh National University
Al-Farabi av. 71
Almaty, 050078, Kazakhstan
kaltayev@kazsu.kz

Q. Zhang
Institute of Aerodynamics
RWTH Aachen

Wuelnnerstr. zw. 5 u.7
Aachen, 52062, Germany
office@aia.rwth-aachen.de

U.K. Zhapbasbaev
al-Farabi Kazakh National University
Al-Farabi av. 71
Almaty, 050078, Kazakhstan
nich7@kazsu.kz

Solution of Maxwell's equations on partially unstructured meshes

Yu.I. Shokin, A.S. Lebedev, O.V. Shtyrina, M.P. Fedoruk

Institute of Computational Technologies SB RAS, Lavrentiev Ave. 6, 630090 Novosibirsk, Russia ict@ict.nsc.ru, sasa@ict.nsc.ru, shtyrinaov@ngs.ru, mife@ict.nsc.ru

1 Introduction

The objective of this article is the development of advanced first-principles simulation tools for modelling electromagnetic wave propagation in fibre-based and planar photonic microstructures with complex geometry.

Two major approaches in numerical modelling of photonic devices become de facto standards in industrial and academic research: finite difference time domain (FDTD) methods and finite element methods (FEM).

Most of numerical schemes currently used for complex photonic structures modelling are based on a dual grid with edge-based discretisation and a staggering of the electric and magnetic fields in space and time, and can all be described as generalizations of the original Yee algorithm [1]. The classical finite-difference time-domain method (see e.g. [2] -[4]) based on this algorithm has limitations when the material parameters change rapidly as is the case for the fibre gratings and planar photonic microstructures. Another problem typically associated with FDTD is complicated treatment of complex boundaries. Curved surfaces are usually modelled using a staircase approximation; and to reduce discretisation error, the mesh must be highly refined.

The finite element method is usually used for solving quasi-stationary Maxwell's equations, see e.g. [5], [6]. Typically one is interested in finding spectral parameters of the devices with a very high accuracy. This leads to the solution of eigenvalue problems for the Helmholtz equations.

The limitations on the accuracy, scalability, and solution time make further advances using these approaches increasingly difficult.

We propose a new robust and geometrically flexible finite-volume time domain method (FVTD) for numerical design and characterization of photonic devices. The examples of such devices include fibre tapers, novel class of fibre grating structures with complicated geometries, laser-written silica-on-silicon planar structures and many others.

2 Numerical method

In the absence of charge and current nondimensional Maxwell's equations read as

$$\frac{\partial \mathbf{D}}{\partial t} - \mathrm{rot}\mathbf{H} = 0, \quad \mathbf{D} = \varepsilon \mathbf{E}, \tag{1}$$

$$\frac{\partial \mathbf{B}}{\partial t} + \mathrm{rot}\mathbf{E} = 0, \quad \mathbf{B} = \mu \mathbf{H}, \tag{2}$$

$$\mathrm{div}\mathbf{D} = 0,$$

$$\mathrm{div}\mathbf{B} = 0,$$

where $\mathbf{E}$ and $\mathbf{H}$ — electric and magnetic field, $\mathbf{D}$ and $\mathbf{B}$ — electric and magnetic induction, ε — electric permittivity, μ — magnetic permeability. In this paper we set $\mu = 1$, so that $\mathbf{B} = \mathbf{H}$.

Maxwell's equations (1), (2) in cylindrical coordinates (z, r, θ) take the form

$$\varepsilon \frac{\partial}{\partial t} E_z - \frac{1}{r}\frac{\partial}{\partial r} r H_\theta + \frac{1}{r}\frac{\partial}{\partial \theta} H_r = 0,$$

$$\varepsilon \frac{\partial}{\partial t} E_r - \frac{1}{r}\frac{\partial}{\partial \theta} H_z + \frac{\partial}{\partial z} H_\theta = 0,$$

$$\varepsilon \frac{\partial}{\partial t} E_\theta - \frac{\partial}{\partial z} H_r + \frac{\partial}{\partial r} H_z = 0,$$

$$\frac{\partial}{\partial t} H_z + \frac{1}{r}\frac{\partial}{\partial r} r E_\theta - \frac{1}{r}\frac{\partial}{\partial \theta} E_r = 0, \tag{3}$$

$$\frac{\partial}{\partial t} H_r + \frac{1}{r}\frac{\partial}{\partial \theta} E_z - \frac{\partial}{\partial z} E_\theta = 0,$$

$$\frac{\partial}{\partial t} H_\theta + \frac{\partial}{\partial z} E_r - \frac{\partial}{\partial r} E_z = 0.$$

Description of numerical method will be given for an axially symmetric case with longitudinal and radial components of electric field and azimuthal component of magnetic field equal identically to zero: $D_z = D_r = H_\theta = 0$. In this case Maxwell's equations describe a so called TM-mode and for the dependent variables D_θ, H_z, H_r take the following form

$$\frac{\partial}{\partial t} D_\theta - \frac{\partial}{\partial z} H_r + \frac{\partial}{\partial r} H_z = 0,$$

$$\frac{\partial}{\partial t} r H_z + \frac{\partial}{\partial r} \frac{r D_\theta}{\varepsilon} = 0, \tag{4}$$

$$\frac{\partial}{\partial t} r H_r - \frac{\partial}{\partial z} \frac{r D_\theta}{\varepsilon} = 0.$$

Introducing the notations

$$\mathbf{U} = \begin{pmatrix} U_1 \\ U_2 \\ U_3 \end{pmatrix} = \begin{pmatrix} D_\theta \\ rH_z \\ rH_r \end{pmatrix}, \quad \mathbf{V} = \begin{pmatrix} V_1 \\ V_2 \\ V_3 \end{pmatrix} = \begin{pmatrix} E_\theta \\ H_z \\ H_r \end{pmatrix}, \quad \mathbf{U} = \begin{pmatrix} \varepsilon & 0 & 0 \\ 0 & r & 0 \\ 0 & 0 & r \end{pmatrix} \mathbf{V} \quad (5)$$

and using (x_1, x_2) along with (z, r) for independent variables, one can rewrite the equations (4) in the vector form

$$\frac{\partial}{\partial t} \mathbf{U} + \frac{\partial}{\partial x_1} A_1 \mathbf{U} + \frac{\partial}{\partial x_2} A_2 \mathbf{U} = 0, \quad (6)$$

where

$$A_1 = \begin{pmatrix} 0 & 0 & -1/r \\ 0 & 0 & 0 \\ -r/\varepsilon & 0 & 0 \end{pmatrix}, \quad A_2 = \begin{pmatrix} 0 & 1/r & 0 \\ r/\varepsilon & 0 & 0 \\ 0 & 0 & 0 \end{pmatrix}.$$

At this point we suppose that a computational domain is covered with a triangular grid. By integrating the equation (6) over the triangular cell Δ_i and applying Gauss's theorem to transform integrals with spatial derivatives, we get

$$\frac{\partial}{\partial t} \int_{\Delta_i} \mathbf{U} dz dr + \sum_{k=1}^{3} \int_{\Gamma_i^k} A \mathbf{U} d\Gamma = 0, \quad (7)$$

where Γ_i^k are the edges of the cell, $\mathbf{n} = (n_1, n_2)$ means the outwards directed unit normal to the corresponding edge,

$$A = n_1 A_1 + n_2 A_2 = \begin{pmatrix} 0 & n_2/r & -n_1/r \\ rn_2/\varepsilon & 0 & 0 \\ -rn_1/\varepsilon & 0 & 0 \end{pmatrix}. \quad (8)$$

The unknowns $\mathbf{U}_i$ associated with the triangle Δ_i are supposed to be defined in the point having coordinates

$$x_{i1}^b = \frac{1}{r_i^m S_i} \int_{\Delta_i} rz dz dr, \quad x_{i2}^b = \frac{1}{r_i^m S_i} \int_{\Delta_i} r^2 dz dr,$$

where (z_i^m, r_i^m) and S_i are the median intersection point and the area of the triangle correspondingly. The point $\mathbf{x}_i^b = (x_{i1}^b, x_{i2}^b)$ is situated in the gravity center of the triangular element taking account three-dimensionality of the element (remind that axially symmetrical case is considered). It is easy to prove that mean value over the cell Δ_i of an arbitrary sufficiently smooth function $f(z, r)$ equals to second order accuracy to the value of the function at $(z_i^b, r_i^b) = (x_{i1}^b, x_{i2}^b)$, i.e.

$$\frac{1}{r_i^m S_i} \int_{\Delta_i} rf(z, r) dz dr = f(z_i^b, r_i^b) + O(\delta^2),$$

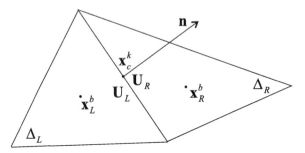

Fig. 1. Two neighbouring triangular cells Δ_L and Δ_R with barycenters $\mathbf{x}_L^b$ and $\mathbf{x}_R^b$ correspondingly; the midpoint of the common edge is denoted by $\mathbf{x}_c^k$

where δ is the size of the grid element. Note that in plane case the unknowns $\mathbf{U}_i$ are defined at the median intersection point, i.e. $(x_{i1}^b, x_{i2}^b) = (x_{i1}^m, x_{i2}^m)$.

Let's $\mathbf{U}^n$ denote the vector of unknowns defined at element barycenter $\mathbf{x}^b$ at time $t_n = n\tau$, where τ is the time step. Approximating in (7) the time derivative by finite difference, for the i-th cell we get

$$S_i \frac{\mathbf{U}_i^{n+1} - \mathbf{U}_i^n}{\tau} + \sum_{k=1}^{3} \int_{\Gamma_i^k} A\mathbf{U} d\Gamma = 0. \tag{9}$$

To find $\mathbf{U}_i^{n+1}$ from (9), one have to evaluate the integrals of the flux $A\mathbf{U}$ over the edges of the triangle Δ_i. Denoting an approximate value of the flux at the center of the k-th edge by $\mathbf{F}_k$ and the length of the k-th edge by l_k, and applying the midpoint quadrature formula, we have

$$\mathbf{U}_i^{n+1} = \mathbf{U}_i^n - \frac{\tau}{S_i} \sum_{k=1}^{3} l_k \mathbf{F}_k. \tag{10}$$

2.1 Computation of fluxes

Suppose that in the triangle Δ_L functions $\mathbf{U}$ vary linearly. Using values $\mathbf{U}$ in the barycenter of the triangle and gradients of $\mathbf{U}$ in the triangle, one can find the functions at the center of an edge which are denoted by $\mathbf{U}_L$. Analogously, $\mathbf{U}_R$ are the functions at the center of the same edge computed from adjacent triangle Δ_R. Using $\mathbf{U}_L$ and $\mathbf{U}_R$, in general different, we find the flux $\mathbf{F}_k$ through the center of the edge Γ^k separating the triangles Δ_L and Δ_R.

If matrix A is independent of ε or if coefficient ε have equal values for triangles Δ_L and Δ_R then the flux $\mathbf{F}_k$ is computed by solving one-dimensional, with respect to the normal to the edge, Riemann problem for equation

$$\frac{\partial}{\partial t}\mathbf{U} + A\frac{\partial}{\partial n}\mathbf{U} = 0,$$

and is given by the formula

$$\mathbf{F}_k = A^+ \mathbf{U}_L + A^- \mathbf{U}_R, \tag{11}$$

where

$$A^\pm = R^{-1} diag \left\{ \frac{1}{2} \left(\lambda_k \pm |\lambda_k| \right) \right\} R,$$

R is the matrix whose columns are the right eigenvectors of matrix A, and λ_k are eigenvalues of A. For matrix A given by (8) we have

$$A^+ = \frac{1}{2} \begin{pmatrix} \dfrac{1}{\sqrt{\varepsilon}} & \dfrac{n_2}{r} & -\dfrac{n_1}{r} \\[2mm] \dfrac{rn_2}{\varepsilon} & \dfrac{n_2^2}{\sqrt{\varepsilon}} & -\dfrac{n_1 n_2}{\sqrt{\varepsilon}} \\[2mm] -\dfrac{rn_1}{\varepsilon} & -\dfrac{n_1 n_2}{\sqrt{\varepsilon}} & \dfrac{n_1^2}{\sqrt{\varepsilon}} \end{pmatrix}, \quad A^- = \frac{1}{2} \begin{pmatrix} -\dfrac{1}{\sqrt{\varepsilon}} & \dfrac{n_2}{r} & -\dfrac{n_1}{r} \\[2mm] \dfrac{rn_2}{\varepsilon} & -\dfrac{n_2^2}{\sqrt{\varepsilon}} & \dfrac{n_1 n_2}{\sqrt{\varepsilon}} \\[2mm] -\dfrac{rn_1}{\varepsilon} & \dfrac{n_1 n_2}{\sqrt{\varepsilon}} & -\dfrac{n_1^2}{\sqrt{\varepsilon}} \end{pmatrix}.$$

When coefficient ε varies within the computational domain, especially when ε have jump discontinuities, numerical experiments and comparisons with exact solutions confirm the effectiveness of the following way to compute fluxes $\mathbf{F}_k$. First, rewrite the formula (11) for $\mathbf{F}_k$ using the vector $\mathbf{V}$ from (5), the coefficient ε being defined later

$$\mathbf{F}_k = A^+ \mathbf{U}_L + A^- \mathbf{U}_R = A^+ \begin{pmatrix} \varepsilon & 0 & 0 \\ 0 & r & 0 \\ 0 & 0 & r \end{pmatrix} \mathbf{V}_L + A^- \begin{pmatrix} \varepsilon & 0 & 0 \\ 0 & r & 0 \\ 0 & 0 & r \end{pmatrix} \mathbf{V}_R,$$

or

$$\mathbf{F}_k = \begin{pmatrix} 1 & 0 & 0 \\ 0 & r & 0 \\ 0 & 0 & r \end{pmatrix} \left(C^+ \mathbf{V}_L + C^- \mathbf{V}_R \right), \tag{12}$$

where

$$C^+ = \frac{1}{2} \begin{pmatrix} \sqrt{\varepsilon} & n_2 & -n_1 \\[2mm] n_2 & \dfrac{n_2^2}{\sqrt{\varepsilon}} & -\dfrac{n_1 n_2}{\sqrt{\varepsilon}} \\[2mm] -n_1 & -\dfrac{n_1 n_2}{\sqrt{\varepsilon}} & \dfrac{n_1^2}{\sqrt{\varepsilon}} \end{pmatrix}, \quad C^- = \frac{1}{2} \begin{pmatrix} -\sqrt{\varepsilon} & n_2 & -n_1 \\[2mm] n_2 & -\dfrac{n_2^2}{\sqrt{\varepsilon}} & \dfrac{n_1 n_2}{\sqrt{\varepsilon}} \\[2mm] -n_1 & \dfrac{n_1 n_2}{\sqrt{\varepsilon}} & -\dfrac{n_1^2}{\sqrt{\varepsilon}} \end{pmatrix}.$$

Setting $r = r_c^k$ in (12), where r_c^k is the r-coordinate of the center of the edge, and denoting by $\tilde{C}^+$ and $\tilde{C}^-$ the matrices C^+ and C^- with coefficient ε averaged by the formula $\tilde{\varepsilon} = 2\varepsilon_L \varepsilon_R / (\varepsilon_L + \varepsilon_R)$, we arrive at the following formula for $\mathbf{F}_k$

$$\mathbf{F}_k = \begin{pmatrix} 1 & 0 & 0 \\ 0 & r_c^k & 0 \\ 0 & 0 & r_c^k \end{pmatrix} \left(\tilde{C}^+ \mathbf{V}_L + \tilde{C}^- \mathbf{V}_R \right). \tag{13}$$

Note that the matrix in the relation (5) between vectors $\mathbf{V}$ and $\mathbf{U}$ is computed with $\varepsilon = \varepsilon_L$ and $r = r_L^m$ for the vectors at the barycenter of the triangle Δ_L, and with $\varepsilon = \varepsilon_L$ and $r = r_c^k$ for the vectors at the center of the k-th edge.

2.2 Computation of functions at the edge center

To evaluate the fluxes $\mathbf{F}_k$ by the formula (13) which are used in (10) to compute $\mathbf{U}_i^{n+1}$, one have to find the vectors $\mathbf{V}_L$ and $\mathbf{V}_R$. In order for the method to be of the second order, we find the vectors at $t = t_n + \tau/2$. The vector $\mathbf{V}_L^{n+1/2}$ is computed from the triangle Δ_L in the following way (similarly the vector $\mathbf{V}_R^{n+1/2}$ is computed from the triangle Δ_R).

By rewriting the equations (4) in terms of the vector $\mathbf{V}$, we have

$$\frac{\partial}{\partial t}\mathbf{V} + B_1\frac{\partial}{\partial x_1}\mathbf{V} + B_2\frac{\partial}{\partial x_2}\mathbf{V} + \mathbf{Q} = 0, \tag{14}$$

where

$$B_1 = \begin{pmatrix} 0 & 0 & -1/\varepsilon \\ 0 & 0 & 0 \\ -1 & 0 & 0 \end{pmatrix}, \quad B_2 = \begin{pmatrix} 0 & 1/\varepsilon & 0 \\ 1 & 0 & 0 \\ 0 & 0 & 0 \end{pmatrix}, \quad \mathbf{Q} = \begin{pmatrix} 0 \\ V_1/r \\ 0 \end{pmatrix}.$$

Using the theorem of Taylor we get second order approximation to $V_{Lj}^{n+1/2} = V_j\left(\mathbf{x}_c^k, t_n + \tau/2\right)$:

$$V_{Lj}^{n+1/2} = V_j\left(\mathbf{x}_L^b, t_n\right) + \sum_{i=1}^{2}\frac{\partial V_j}{\partial x_i}\left(x_{ci}^k - x_{Li}^b\right) + \frac{\tau}{2}\frac{\partial V_j}{\partial t},$$

or, substituting the time derivative from (14) into the above relation,

$$V_{Lj}^{n+1/2} = V_j\left(\mathbf{x}_L^b, t_n\right) + \sum_{i=1}^{2}\frac{\partial V_j}{\partial x_i}\left(x_{ci}^k - x_{Li}^b\right)$$
$$- \frac{\tau}{2}\left\{\sum_{i=1}^{3}\left(b_{ji}^1\frac{\partial V_i}{\partial x_1} + b_{ji}^2\frac{\partial V_i}{\partial x_2}\right) + Q_j\right\}. \tag{15}$$

In the formula (15) the notations b_{ji}^1 and b_{ji}^2 are used for the elements of the matrices B_1 and B_2 correspondingly, $V_j\left(\mathbf{x}_L^b, t_n\right)$ are known components of the vector $\mathbf{V}$ at the barycenter $\mathbf{x}_L^b$ of the triangle under consideration, and $\partial V_j/\partial x_i$ are partial derivatives which remain to be evaluated.

2.3 Computation of the gradients

In our calculations we estimate partial derivatives $\partial V_j/\partial x_i$ in the same manner for all the components of the vector $\mathbf{V}$. So, to simplify the notation, we

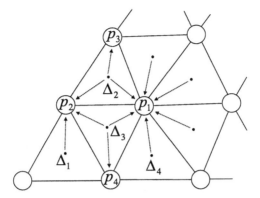

Fig. 2. The gradients are found in two steps starting by extrapolation the values from the barycenters onto the vertexes

describe the procedure for a scalar function f_i defined at the barycenter of the triangle Δ_i.

The procedure involves two stages. At the preliminary stage (Fig. 2) values of the function are extrapolated into the vertexes of the triangles. To extrapolate, we assume that within the triangle Δ_i there is a linear function $u_i(x_1, x_2)$ that takes value f_i at the barycenter of the triangle, and has gradient defined by given values at the barycenters of three triangles adjacent to triangle Δ_i. For example, the gradient of $u_3(x_1, x_2)$ in the triangle Δ_3 (Fig. 2) is defined by values f_1, f_2 and f_4 from barycenters of Δ_1, Δ_2 and Δ_4. Function $u_3(x_1, x_2)$ produces values at the vertexes p_1, p_2 and p_4. So at the point p_1 there are six values in all, according to the number of triangles having the point p_1 as one of their vertexes. Arithmetic mean of the six values is then assigned to the point p_1. Obviously, the vertex p_2 gets the arithmetic mean of three values produced by triangles Δ_1, Δ_2 and Δ_3.

Note that a triangle lying on the boundary of the computational domain has only two neighbours. The gradient for a boundary triangle is computed using values of the function from barycenters of the triangle itself and of its two neighbours. It should be noted that in case of unsound triangulation the three barycenters may form degenerate triangle.

It also may happen that a boundary triangle has only one neighbour. For example, the triangle Δ_1 borders only triangle Δ_3. At the preliminary stage the gradient for the triangle Δ_1 is set equal to the gradient for the triangle Δ_3.

At the second stage final gradients are computed that are used in the formula (15). This time all triangles are treated equally, the gradient for a triangle being defined by the values of a function at three vertexes of the triangle.

Fig. 3. Grid for a region with holes

Such a two-phase algorithm increases the number of point values involved in computation of gradients for a triangular cell, which is especially important, as is proved by numerical experiments, for the boundary triangles.

3 Numerical examples

In this section we present the results of test calculations performed with the numerical algorithm described. The calculations have been conducted in regions of different shapes, including square, circle and regions with holes. In general, to build the grid we start by domain boundary discretization and then use Delaunay type boundary preserving triangulation method with the original smoothing technique. For rectangular regions triangular grids are also used, which we call regular grids. Such grids are build by first dividing a region into quadrilaterals and then dividing every quadrilateral into four triangular cells by diagonals. In the Figure 3 an example of triangular grid with smoothly varying cell size for a region with holes is shown. Cells near the holes are ten times smaller than the cells near the outer boundary.

As an indicator of departure of numerical solution (U_1^n, U_2^n, U_3^n) from exact solution (E_1^n, E_2^n, H_3^n) at $t = t_n$ we use the following quantity

$$\delta = \frac{\sqrt{\sum_i \left[(U_{1i}^n - E_{1i}^n)^2 + (U_{2i}^n - E_{2i}^n)^2 + (U_{3i}^n - H_{3i}^n)^2\right] S_i}}{\sqrt{\sum_i \left[(E_{1i}^n)^2 + (E_{2i}^n)^2 + (H_{3i}^n)^2\right] S_i}} \qquad (16)$$

where the summation is carried out over all triangles, S_i is the area of i-th triangle.

Test 1. Axially symmetric TM-mode.

In the case of axially symmetric TM-mode the Maxwell's equations take the form (4). For piecewise constant ε the system (4) has exact solution

$$0 \le r \le a \quad (\varepsilon = \varepsilon_1): \quad E_\theta = \quad k\frac{a}{u} J_1\left(\frac{u}{a}r\right) \sin\left(\omega t - \beta z\right),$$

$$H_z = \quad J_0\left(\frac{u}{a}r\right) \cos\left(\omega t - \beta z\right),$$

$$H_r = -\beta\frac{a}{u} J_1\left(\frac{u}{a}r\right) \sin\left(\omega t - \beta z\right),$$

$$r > a \quad (\varepsilon = \varepsilon_2): \quad E_\theta = -k\frac{a}{w}\frac{J_0\left(u\right)}{K_0\left(w\right)} K_1\left(\frac{w}{a}r\right) \sin\left(\omega t - \beta z\right),$$

$$H_z = \quad \frac{J_0\left(u\right)}{K_0\left(w\right)} K_0\left(\frac{w}{a}r\right) \cos\left(\omega t - \beta z\right),$$

$$H_r = \quad \beta\frac{a}{w}\frac{J_0\left(u\right)}{K_0\left(w\right)} K_1\left(\frac{w}{a}r\right) \sin\left(\omega t - \beta z\right).$$

The constant β is found by solving the nonlinear equation

$$\frac{\varepsilon_1}{\varepsilon_2}\frac{J_0'(u)}{uJ_0(u)} + \frac{K_0'(u)}{wK_0(u)} = 0 \qquad (17)$$

where $u = a(\varepsilon_1 k^2 - \beta^2)^{1/2}$, $w = a(\varepsilon_2\beta^2 k^2 - \beta^2)^{1/2}$; a is the radius of the core of the fiber, $\sqrt{\varepsilon_1}, \sqrt{\varepsilon_2}$ are the indexes of refraction for the core and cladding correspondingly, k is the wave number, J_0, K_0 are Bessel functions. The following parameters are used: $\omega = k = 2\pi$, $a = 4.0/k$, $\varepsilon_1 = 2.1025$, $\varepsilon_2 = 1.0$. In this case $\beta = 4.944844443409/a$, $u = 3.031256081311$, $w = 2.907144057235$.

Numerical solution to the equations (4) is found by the formulas (10), (11). Table 1 shows maximal errors δ_1 and δ_2 for different grids. The error denoted by δ_1 refers to the case when the gradients in triangles are set equal to zero thus reducing the approximation to the first order. The error δ_2 relates to the second order approximation of functions within a triangle. The errors behaviour is seen to correspond to the order of accuracy.

Ten period evolution of δ_2 error is presented in the Figure 4 for different irregular triangular grids.

Test 2. Hybrid modes in the step-index fibers.

Electromagnetic field for the step-index fiber is described by equations (3) with piecewise constant ε. In this case the equations (3) have exact solution with z-components of the electromagnetic field expressed by (see e.g. [7]):

$$0 \le r \le a \quad (\varepsilon = \varepsilon_1): \quad E_z = \quad J_m\left(\frac{u}{a}r\right) \cos\left(m\theta\right) \cos\left(kt - \beta z\right),$$

$$H_z = -\frac{\beta}{k}sJ_m\left(\frac{u}{a}r\right) \sin\left(m\theta\right) \cos\left(kt - \beta z\right),$$

Table 1. Maximal errors δ_2 and δ_1 for computed TM-mode (cylindrical coordinates, ε has jump discontinuity).

grid type	number of triangles	δ_2	δ_1
regular	$16 \times 16 \times 4$	0.0267	0.337
	$32 \times 32 \times 4$	0.00756	0.190
	$64 \times 64 \times 4$	0.00214	0.0990
	$128 \times 128 \times 4$	0.000867	0.0510
	$256 \times 256 \times 4$	0.000205	0.0259
regular, conforming to the jump of ε	$16 \times 16 \times 4$	0.0265	
	$32 \times 32 \times 4$	0.00726	
	$63 \times 63 \times 4$	0.00208	
	$126 \times 126 \times 4$	0.000643	
	$252 \times 252 \times 4$	0.000211	
irregular	240	0.106	
	948	0.0267	
	3808	0.00797	
	15194	0.00222	
	60608	0.000677	
	242186	0.000218	

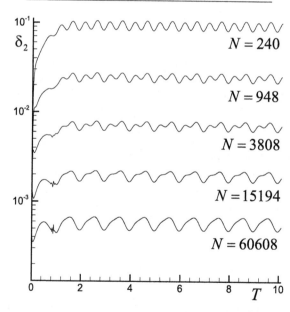

Fig. 4. The error δ_2 for TM-mode (irregular grid, the functions are linear within a cell)

$$r > a \quad (\varepsilon = \varepsilon_2): \quad E_z = \frac{J_m(u)}{K_m(w)} K_m\left(\frac{w}{a} r\right) \cos(m\theta) \cos(kt - \beta z),$$

$$H_z = -\frac{\beta}{k} s \frac{J_m(u)}{K_m(w)} K_m\left(\frac{w}{a} r\right) \sin(m\theta) \cos(kt - \beta z).$$

Here

$$s = m \left(\frac{1}{u^2} + \frac{1}{w^2}\right) \Bigg/ \left(\frac{J_m'(u)}{u J_m(u)} + \frac{K_m'(w)}{w K_m(w)}\right),$$

m is an integer, and other notations were introduced previously. The components E_r, E_θ, H_r, and H_θ can be expressed in terms of E_z and H_z using equations (3).

The propagation constant β of the hybrid modes is calculated by solving the dispersion equation

$$\left[\frac{J_m'(u)}{u J_m(u)} + \frac{K_m'(w)}{w K_m(w)}\right] \left[\frac{J_m'(u)}{u J_m(u)} + \frac{\varepsilon_2}{\varepsilon_1} \frac{K_m'(w)}{w K_m(w)}\right] =$$

$$m^2 \left(\frac{1}{u^2} + \frac{1}{w^2}\right) \left[\frac{1}{u^2} + \frac{\varepsilon_2}{\varepsilon_1} \frac{1}{w^2}\right].$$

The following parameters are used: $\omega = k = 6.0$, $a = 0.64$, $\varepsilon_1 = 2.25$, $\varepsilon_2 = 1.00$ $u = 2.0638374168422707764$, $w = 3.7646480734381830207$, $\beta = 8.4024409232586757340$, $m = 1$.

Figure 5 shows computational domain covered partially unstructured mesh. The cells of the mesh are triangular based prisms. Arrows in the Figure 5 indicate directions of magnetic field $\mathbf{H}$ at $t = 0.46T$.

In the Figure 6 the contour plot for E_z-component at $t = 0.46T$ on the surface of the computational domain lying in the core region ($r \leq a$) is shown.

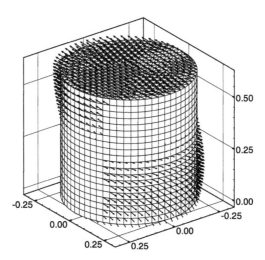

Fig. 5. Computed magnetic field directions for hybrid mode

Figure 7 demonstrates five period evolution of δ_2 error for different grids having 166×10, 676×20, 2718×40, and 10884×80 cells (here for example 166 and 10 are the number of triangles in (r, θ) plain and the number of prisms in longitudinal direction). From the Figure 7 it is seen that two times refinement of the mesh results in four times reduction of the error in accordance with the expected second order of the numerical method.

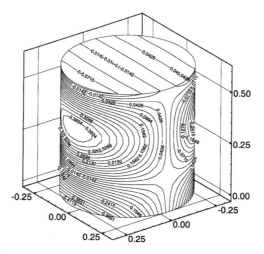

Fig. 6. Computed E_z-component for hybrid mode

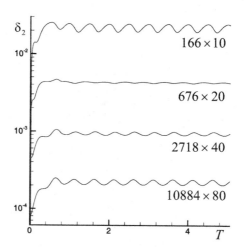

Fig. 7. The error δ_2 for hybrid mode (irregular grid, the functions are linear within a cell)

4 Conclusion

A finite volume method to solve Maxwell's equations on partially unstructured grids for the media having jump discontinuities in the electric permittivity coefficient ε has been presented. Test numerical simulations prove the second order accuracy of the method proposed.

References

1. Yee KS (1966) IEEE Trans Antennas Propagat 17:585–589
2. Taflove A (1995) Computational electrodynamics. The finite-difference time-domain method. Artech House, Boston
3. Taflove A (editor) (1998) Advances in computational electrodynamics. The finite-difference time-domain method. Artech House, Boston
4. Taflove A, Hagness SC (2000) Computational electrodynamics. The finite-difference time-domain method (second edition). Artech House, Boston
5. Fujisawa T, Koshiba M (2003) Optics Express 11:1482–1489
6. Obayya SSA, Rahman BMA, El-Mikati HA (2000) J Lightwave Technol 18:409–415
7. Okamoto K (2000) Fundamentals of optical waveguides. Academic Press, London

The integral equations method in problems of electrical sounding

M. Orunkhanov and B. Mukanova

Institute of Mathematics and Mechanics, Al-Farabi Kazakh National University, Masanchi str. 39/47, 050012 Almaty, Kazakhstan
orun@kazsu.kz, mubaga@kazsu.kz

Summary. The apparent resistivity for vertical profiling above local heterogeneity is calculated. The problem is solved by integral equations method. The vertical profiling inverse problem is formulated and some numerical calculations methods are suggested.

Introduction

We consider the problem on vertical profiling above a local heterogeneity by the method of resistances. The measuring equipment consists of a source probe, which provides direct current flows down to the researched environment and a series of pickup electrodes.

The apparent resistivity is calculated by a potential difference and a current value. As a result we obtain a sounding curve, which depends on the apparent resistivity at the point of measurement. The distances between pickup electrodes are neglected in theoretical calculations([1]). Hence the apparent resistivity can be considered as a derivative of an electric field with respect to the measuring line.

In the method of vertical profiling, the measuring equipment is moving as a whole along a sounding line. A series of curves of apparent resistivities obtained for each position of the measuring equipment are the main material for analysis.

1 Mathematical model

Let us formulate the mathematical model of a vertical profiling problem. We assume, that environment is not electrically uniform and has two-dimensional piecewise constant distribution of conductivity. Concerning the geometry of bed layers we make the following assumptions:

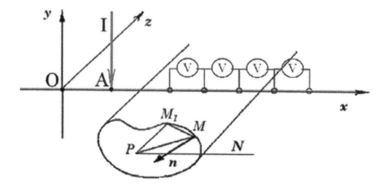

Fig. 1. The scheme of vertical electrical sounding and the geometry of attitude of beds

1. Let the plane (x, z) in the Cartesian coordinates to coincide with a surface of the Earth, and the distribution of conductivities to be independent of a coordinate z.

2. A parameterization exists for a cross-section of the boundary Γ by a plane $z = const$ in the following polar coordinates: $r = r(\theta)$, with the center in some point $P(x_P, y_P)$ (Fig. 1). The section of the boundary forms a closed curve in a plane (x, y) and the function $r = r(\theta)$ satisfies the conditions:

$$r(\theta) \in C^2([0, 2\pi]), 0 < R_1 \le r(\theta) \le R_2, \tag{1}$$
$$\max(|r'(\theta)|, |r''(\theta)|) \le K, m = R_1 - 2\pi K > 0, \theta \in [0, 2\pi).$$

Though the distribution of heterogeneity is two - dimensional, the field parameters depend on three space coordinates due to the fact that the electric field in the environment is raised by a point source. For a stationary field and in absence of volume sources the electrostatic potential is defined the by Laplace's equation in points of a domain:

$$\Delta\varphi = 0, \tag{2}$$

with the following boundary conditions:

$$\varphi|_{\Gamma-} = \varphi|_{\Gamma+}, \quad \sigma_1 \frac{\partial\varphi}{\partial n}\bigg|_{\Gamma-} = \sigma_2 \frac{\partial\varphi}{\partial n}\bigg|_{\Gamma+}, \tag{3}$$

The conditions of decreasing on infinity $\varphi(\infty) = 0$ and a boundary condition on the Earth surface should be satisfied as well as:

$$\frac{\partial\varphi}{\partial y}\bigg|_{y=0} = -I(\delta(\overrightarrow{r} - \overrightarrow{OA})). \tag{4}$$

Let the solution of the problem be the sum of potentials of a point source in homogeneous half-space and the unknown regular additive:

$$\varphi = U_0(M) + u(M) = \frac{I}{2\sigma_1\pi|MA|} + u(M). \tag{5}$$

Also let the function $u(M)$ satisfy the Laplace's equation everywhere, except the geoelectrical boundary Γ, and the boundary conditions (4-5) be:

$$\sigma_1\frac{\partial u}{\partial n}\Big|_{\Gamma_+} - \sigma_2\frac{\partial u}{\partial n}\Big|_{\Gamma_-} = -\left(\sigma_1\frac{\partial U_0}{\partial n}\Big|_{\Gamma_+} - \sigma_2\frac{\partial U_0}{\partial n}\Big|_{\Gamma_-}\right), \tag{6}$$

$$\frac{\partial u}{\partial y}\Big|_{y=0} = 0. \tag{7}$$

Let us search the solution $u(M)$ as a potential of a simple fiber created by secondary charges, distributed on the geoelectric boundary Γ, and its mapping in the upper half-space. The symmetric mapping is used to provide the condition (7) on a surface of the Earth. The density of a simple fiber $\nu(M)$ is considered as a required function.

The unknown density $\nu(M)$ should satisfy the integral equation obtained from the conditions on a normal derivative of a field on Γ and the Green's formulas:

$$\nu(M) = \frac{\lambda}{2\pi}\iint_\Gamma \nu(M_1)\frac{\partial}{\partial n}\left(\frac{1}{r_{MM_1}} + \frac{1}{r_{MM_1'}}\right)d\Gamma(M_1) + \lambda F_0(M), \tag{8}$$

where $F_0(M) = \partial U_0/\partial n(M)$, $\lambda = (\sigma_1 - \sigma_2)/(\sigma_1 + \sigma_2)$.

Here the points M, M_1 belong to a surface of integration Γ, and M_1' - to its mapping in the upper half-space, r_{MM_1}, r'_{MM_1} are the distances from the point M to M_1 and M_1' respectively.

In distinction from the equation for the sounding problem above an inclined seam [2], we should consider the additional term in (8), because the boundary of heterogeneity is not a plane in this case. The equation has the singularity because the points M, M_1 belong to the same surface.

In [3] we have obtained the conditions of solvability of the equation (8). It is turned out, that the following iterative scheme converges in norm of C under the condition of the second order smoothness of the boundary and under some restrictions on the value of derivatives in the equation of the surface:

$$\nu_{n+1}(M) = \frac{\lambda}{2\pi}\iint_\Gamma \nu_n(M_1)\frac{\partial}{\partial n}\left(\frac{1}{r_{MM_1}} + \frac{1}{r_{MM_1'}}\right)d\Gamma(M_1) + \lambda F_0(M), \tag{9}$$

$$|\nu_{n+1} - \nu_n|_C \le |\nu_n - \nu_{n-1}|_C \cdot \frac{\lambda}{2\pi}(C_1 + C\pi), \qquad n = 0,1,2,... \tag{10}$$

We specify some initial approximation of the function $\nu_0(M) \in C(\Gamma)$. Each next approximation $\nu_{n+1}(M)$ can be calculated from the expression (9).

It was proved in [3], that the iterative process converges as a geometric series with a geometric ratio $\lambda(C_1+C\pi)/2\pi$, where the constants C, C_1 depend on K, m, R_1, R_2 from the condition (1).

2 Numerical results

We realized numerically the algorithm described above and developed the adequate visual interface. The program allows the user to set independently the form of inclusion, parameters of heterogeneity and the positions of source probes.

In order to set the embedding profile, the user independently renders reference points on which the program carries out either a spline interpolation or a Fourier approximation and renders a section contour. Then the user launchs a calculation window, where it is also possible to set the following parameters of calculations: the maximum of acceptable number of iterations and the calculation accuracy for $\nu_n(M)$. The program calculates the apparent resistivities and put these curves on a graphic schemes along a vertical axis under each source probe in accordance with the data order accepted in geological practice.

In spite of the fact that the field has a three-dimensional character, the computing capability of the integral equation method are comparable with two-dimensional problems.

Having obtained some experience of calculations the user can make the conclusion concerning a heterogeneity's geometry. The program gives the possibility to use the sounding curves for different positions of sounding lines simultaneously.

The calculations with different ratios of medium conductivities and different geometries of heterogeneity show us that the apparent resistivities are weakly sensitive to an increase of the contrast in the case of big contrasts σ_2/σ_1. It is due to the fact that the sounding curves should be situated between two extreme cases - for a conductive embedding with $\sigma_2 \to \infty$, $\lambda = -1$, and for a dielectric embedding with $\sigma_2 \to 0$, $\lambda = +1$. It is possible to demonstrate that the curves collect near the curves with the extreme values $\lambda = \pm 1$ under the condition of increasing σ_1/σ_2 or σ_2/σ_1.

Hence the main factor of the data interpretation problem is a geometry of embedding for such kind of mediums, and not the values of conductivities. It has allowed us to get the next formulation of the vertical profiling inverse problem:

Let us minimize the following functional by varying the function $r(\theta)$:

$$J(r(\theta)) = \sum_{j=1}^{N_{pol}} \left\| \rho_0^j(s) - s^2 \frac{\partial u_j}{\partial s}(s) \right\|_{C[s_{min}, s_{max}]} \longrightarrow min, \qquad (11)$$

$$u_j(s) = u_j(x_A + s\cos\beta, 0, z_A + s\sin\beta), \quad s \in [s_{min}, s_{max}], \quad j = 1, ..., N_{pol},$$

$$u_j(P) = \frac{1}{2\pi} \iint_\Gamma \nu_j(M) \left[\frac{1}{r_{PM}} + \frac{1}{r_{PM'}} \right] d\Gamma(M),$$

$$\Gamma = \left\{ x, y, z \,\middle|\, x = r(\theta)\cos\theta + x_P, y = r(\theta)\sin\theta + y_P, \right.$$

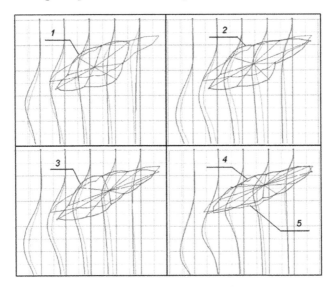

Fig. 2. The reconstruction of the initial profile by a trial method. Curves 1-4 – are consecutive iterations, 5 – required profile, $\sigma_1/\sigma_2 = 100$

$$z \in (-\infty, +\infty), \theta \in [0, 2\pi], r(0) = r(2\pi) \Big\}.$$

Here $\rho^j{}_0(s)$ is an experimental sounding curve for each of j positions of source probe and $\nu_j(M)$ is a solution of the integral equation (8), in which the function $F_0(M)$ is calculated for each of j source probe position. N_{pol} is a number of different locations of source probes. The coefficient $\lambda = (\sigma_1 - \sigma_2)/(\sigma_2 + \sigma_1)$ is assumed to be known.

A trial method is the first step on the way of solving the problem. On the figure 2 there are the results of successful linear search of embedding profile by the curve family which has been reached after 4 iterations. For comparison the initial contour also is shown on the figure. The method was found to be very effective for a simple form of embedding.

In order to solve the problem (11), it is necessary to define a gradient of functional. In the numerical realization of (11) we limit a problem by the discrete approximation of it, because the direct definition of a gradient is a great technical difficulty.

We correspond basic points $M_1, M_2, .., M_n$ to the curve $r(\theta)$, so it is possible to restore the section profile.

Then the functional (11) depends on a finite point coordinate number, and its gradient can be approximated as:

$$\frac{\partial J}{\partial r_k} \approx \frac{J(M_1, M_2, .., M_k + \Delta_k, .., M_n) - J(M_1, M_2, .., M_K, .., M_n)}{\Delta_k}, \quad (12)$$

in every point M_k.

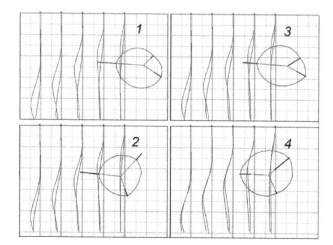

Fig. 3. The partly automated reconstruction of the initial profile with three base points. The numbers on the figures are iteration numbers

There is the movement of the point M_k of a contour along own radius in $M_k + \Delta_k$. We move this point slightly, and therefore we obtain the deformation of a contour in the direction of movement We solve the direct problem (2)-(7) for removing each basic point, then we calculate the functional (11) and the gradient components (12) approximately.

Also we partly automated the process of data interpretation on the small number of basic points. The user renders basic points on the working area of window, where the program calculates and indicates the directions of subsequent displacement of points. These directions are shown as the segments drawn from point M_k, which have a module proportional to the module of $\partial J / \partial r_k$. If a value of $\partial J / \partial r_k$ is positive, then the segment is drawn along the radius of M_k, else it is drawn in the opposite direction. On the figure 3 there are consecutive positions of embedding profile and necessary directions for moving basic points.

Every next iteration gives a better approximation to initial sounding curves. But after some iterations the further improvement will be impossible due to a small number of basic points.

This process is automated. In the next figure 4 the calculation results with all of intermediate positions are shown. The gradient decreasing method is used in these calculations. The application of the described method for large basic point number is less effective than the trial method.

From our point of view the integral equations method is the most adequate method for the problems with a sharply defined geoelectrical boundaries. It has following advantages:

-   the method has a high accuracy of calculation of an electrical field and its derivatives;

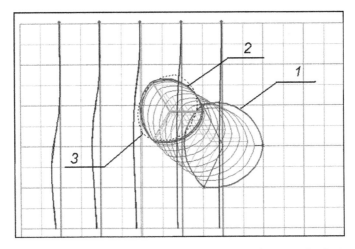

Fig. 4. The reconstruction of initial profile by the gradient method. 1 – the first approximation, 2 – the last approximation, 3 – the required profile, $\sigma_1/\sigma_2 = 100$

- the method is available for the boundaries with complex geometry;
- computing time is comparable to the time of calculation for two-dimensional problems, though a considered electric field is three-dimensional;
- the method is available for three-dimensional geometry of the beds.

We expect also that the method can be implemented in the case of the relief type of the geoelectric boundary and that it can be implemented when the heterogeneity has output area on the Earth surface.

References

1. Khmelevski VK (ed)(1989) Electrical sounding: geophysicist's handbook. Nedra, Moscow (in Russian)
2. Tikhonov AN (1946) On electric sounding above inclined bed. In: Proc. of Institute of Theor. Geophys. Publ. House of AS USSR, Moscow-Leningrad 1:116–136 (in Russian)
3. Orunkhanov M, Mukanova B, Sarbassova B (2004) Comput Technol 9(6):68–72 (in Russian)

The chain of abstraction in High Performance Computing and simulation

M.M. Resch

High Performance Computing Center Stuttgart (HLRS), University of Stuttgart, Nobelstraße 19, 70569 Stuttgart, Germany resch@hlrs.de

Summary. Simulation on Supercomputers has become a standard way of getting insight both in research and in industrial applications like product design and engineering. The process of computing, however, is only a small part of a chain of steps that lead us from reality to simulation results. In recent years the level of performance has dramatically increased and has forced us to reconsider the foundations of supercomputing. In this paper we set out to discuss the state of the art in supercomputing, the theoretical foundations of simulation and the implications that modern supercomputing architectures have on these foundations.

1 Introduction

Simulation has become one pillar of scientific research in the last years [1]. Much as experimental work has changed scientific thinking when it complemented reasoning and logical thought, simulation has started to change the way science works. And just like experiments have not replaced but complemented logic, simulation will not replace any of the other two methods but will complement them. There are a number of problems that come with this. Many simulations can not be verified by experiments and thus have to be taken at face value. Furthermore working in the field of simulation we are aware of the limitations of our physical and mathematical models and know that we have to improve them in order to get better and more reliable simulation results.

When we talk about supercomputing we typically consider it as defined by the TOP500 list [2] or the HPC Challenge benchmark [3]. These lists, however, mainly summarize the fastest systems in terms of some predefined benchmarks. A clear definition of supercomputers is not given. For this article we define the purpose of supercomputing as follows:

- We want to use the fastest system available to get insight that we could not get with slower systems. The emphasis is on getting insight rather than on achieving a certain level of speed.

- We are willing (and able) to spend the necessary amount of money to receive a much higher level of performance than any available standard system can achieve. We accept that this only puts us about 10 years ahead of standard systems.
- We recognize that in order to harvest the potential of such supercomputers we need software. This includes operating systems, programming languages, compilers, parallel programming models, numerical algorithms, libraries, and visualization software.

Any system (hardware and software) that helps to achieve these goals and fulfils the criteria given is considered to be a supercomputer. The definition itself implies that supercomputing and simulations are a third pillar of scientific research and development. Very often simulation complements experiments. To a growing extent, however, supercomputing has reached a point where it can provide insight that can not be achieved using experimental facilities. Some of the fields where this happens are climate research, particle physics or astrophysics. Supercomputing in these fields becomes a key technology if not the only possible one to achieve further breakthroughs.

While the importance of supercomputing increases hardware development causes major changes in our research approach. The increasing performance on the one hand and the new concept of massive parallelism make it necessary to rethink our basic approaches towards simulation.

2 The classic chain of abstraction in simulation

Computer simulation is only one step of abstraction in a rather complex setting. Simulation typically starts with one aspect of reality that is of interest. It aims to simulate that aspect of reality to such an extent that the results of the simulation somehow reflect reality as a whole. By no means does simulation aim to predict reality or explain reality. It only aims to contribute to our better understanding of it. This has to be kept in mind when discussing the following issues.

Given that we use simulation to better understand reality we always first start with reality. There is a chain of abstractions through which we have to go then in order to transform our understanding of reality into a machine readable problem which can then be solved and visualized. The basic steps are shown in Fig. 1.

Starting from reality a physical model has to be developed to describe the process under investigation. Many typical physical models go back to Newtonian physics and are still in use today. They take into consideration the most important physical phenomena related to the problem but still are only an abstract description of reality. The physical model has to be transformed into a mathematical model. This results in systems of equations describing physical phenomena like the flow of a fluid (Navier-Stokes equations) or electrodynamics (Maxwell equations).

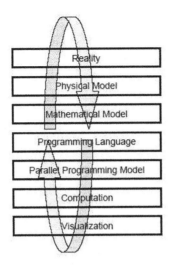

Fig. 1. Layers of abstraction in the simulation cycle

Having created a mathematical abstraction of reality we can tackle it with all the tools available in mathematics since centuries. Analytical solutions would allow us to easily modify parameters in our problem description and immediately get changes of results. A direct interaction between experiments and computational solution would be possible. However, analytical solutions are usually not possible and approximation methods are necessary. This leads us to numerical methods – yet another level of abstraction that is introduced. Numerical methods come with all the well known problems of approximation. Simulation has to be aware of this in order to correctly interpret the results.

Numerical models have to be transformed into computer readable instructions. This is traditionally done using programming languages. FORTRAN was the first high level language to become popular. Its basic concepts still reflect the necessity to describe mathematical problems. Modern programming languages like C++ or Java are based on an object oriented approach. This paradigm follows a more natural approach for expressing ideas and methods in terms of a programming language. However, with the ease of programming comes a loss of performance since these languages ignore the very linear sequential concept of processor architectures.

Finally having expressed the numerical scheme in a machine readable program we can run a simulation making use of a computer. This creates a substantial amount of output data which has to be visualized to be understandable for a human being.Once the results are available they have to be compared with reality to make sure that through all the levels of abstraction not too much information has been lost.

3 Modern supercomputing architectures

Classic supercomputers traditionally were systems that distinguished themselves from standard systems in two ways:

- They were much faster than standard systems
- They were more complex to program

This made it on the one hand more difficult to harvest the potential of a supercomputer. But on the other hand there was a clear advantage for the user in terms of application performance. The outstanding feature of a traditional supercomputer, however, was that it was not much different from a standard system – it was only much faster. The basic concepts of programming could be followed. The same programming languages were used, and with the introduction of Unix even the same operating system was available across the board of systems. The chain of abstraction could work as designed for tens of years. The work invested on standard systems could be carried over to supercomputers and vice versa.

This changed dramatically with the advent of massively parallel systems. As long as the numbers of parallel systems were in the range of tens there were extensions to classical programming languages that helped to bridge the gap. OpenMP [4] is one of these language extensions, which has become popular and quite successful.

For larger number of processors Message Passing Interface (MPI) [5, 6] was defined as a library of calls to explicitly transfer data between processors. MPI became a de facto standard for programming large parallel systems. At the same time that MPI became ubiquitous, standard processors reached a level of performance that matched that of traditional supercomputer.

Clusters turned out to be quite successful for a while. In 2005 cluster installations constituted a large fraction of the fastest systems in the top 500 list [2]. Together with the concept of Grid computing [7] they are considered to form the backbone of national and international supercomputing infrastructures [8]. A number of problems come with the advent of cluster systems [9] that were mostly ignored since for a small number of applications clusters proved to be feasible resources.

Standard processors, however, proved to run into the same physical limitations as supercomputer processors. Clock frequency could not be increased at the rate it was done in the last 20 years. Heat problems made it increasingly difficult to cool systems. Since the number of transistors on the die still could be increased, manufacturers turned to the new concept of "multicores" [10, 11]. With two and more processor cores on a chip the performance can still be increased while the power consumption and heat dissipation remains still at a level that can be controlled.

Modern supercomputing architectures thus become increasingly complex. A single processor is based on multiple cores which typically share caches. A number of such processors are bundled in a shared memory node. A number of such nodes are then connected via some standard network to form a cluster. For the programmer this implies two things:

- A substantial amount of theoretical peak performance in the range of potentially hundreds of Teraflops is available
- The complexity of the system has dramatically increased with all the consequences for stability, programmability, maintainability, and operability.

There are two notable exceptions to this basic concept. On the one hand there is the BlueGene architecture [12] from IBM. It uses slower components to better handle the problem of cooling and imbalance of speed. As a consequence it better overcomes the problems of bandwidth (both for memory and internal network) but on the other hand it requires the user to program hundreds of thousands of processors, thus increasing the level of complexity by orders of magnitudes. On the other hand companies like Cray and NEC still follow the vector architecture approach. This results in systems with a smaller number of processors. The overall system complexity is smaller but the processors itself is more difficult to program. As these architectures show excellent performance for a number of applications but less performance for many others programmability is still a huge issue. A lot of work has to be invested to make such architectures useful for a wide variety of applications [13].

4 The barriers we hit

The new supercomputer architectures provide a level of performance in the range of Teraflops. As a consequence simulation on supercomputing has become a standard tool in a large number of application fields. At the same time we are today able to perform simulations that can be characterized as follows:

- The level of detail in simulations is such that realistic scenarios can be simulated. Simulation thus becomes a tool of everyday research and development.
- The performance level and the size of main memories allow to simulate things that can not be done in experiments. An example for this is atomistic simulation of material behaviour.

In principle this should be positive because it widens the field of application for which simulation can be used. Furthermore, by introducing supercomputers that can do more realistic simulations, simulation becomes a standard tool like experiments. However, at the same time there are some barriers that we hit with our current architectures and processes.

4.1 Reproducibility

When supercomputer experts talk about standardization they mainly mean standards for programming languages or other software tools. These are important. However, standardization is required in a completely different field. Simulation is supposed to complement experiments. The key features of an experiment are the following:

- The parameters and boundary conditions of the experiment are well documented
- The experiment is reproducible and can be conducted by anyone who has the same facilities available

For supercomputer simulations these two criteria do not hold. The basis for computational experiments are codes. These codes are typically well documented and described. However, they can not be described either in such detail or sufficiently abstract to fulfil the requirements for an experiment. Numerical methods are well standardized; however, programming techniques are not. And since programming techniques have an impact on simulation results a lack in describing them is a critical issue for any simulation.

Reproducibility is another feature that can not be guaranteed with supercomputer simulation. If we leave aside that only a few user groups have access to high end systems we still are left with the fact that hardly ever can we guarantee that a large scale system is in a well defined state. Typical simulations are done on systems that are shared with other users. Even if this is not the case we can not guarantee that auxiliary hardware and software (disk systems, file systems, ...) have the same status.

4.2 Verification

When we look at the levels of abstraction in figure 1 we see that the final abstraction layer is that of visualization. Typically a simulation creates a huge amount of data that have to be converted into pictures in order to be understandable for a human being. The circulating arrow in the figure indicates the probably most important process in the simulation chain. The user has to grasp the content of the simulation results and has to link it back to the reality that he intended to describe in the simulation.

For very simple problems this process is a simple one. A driven cavity problem (a typical problem to verify the correctness of flow simulation codes) can easily be studied and the correctness of any solution can be verified – simply because driven cavity is a comparatively simple problem. For more complex problems verification is much more difficult [14] and a lot of work still has to be invested.

A lot of scientific work has been invested in code verification. However, code verification only tackles part of the problem. In our chain of abstractions problems and errors may occur in any step. Code verification can only assure that what was intended to be coded actually was correctly coded. By no means can code verification guarantee that the mathematical model, the numerical method, and the simulation were done correctly. Without verifying the correctness of these steps, however, code verification alone is only a necessary condition but not a sufficient one.

4.3 Visualization & understanding

Visualization is a prerequisite to understand the results of a simulation. At least to some extent visualization also allows us to verify simulation results. However, the growing speed of supercomputers and the growing size of main memories have created a visualization crisis. Over time we have gone through several steps in visualization. Starting with functions to show the correlation of two factors, we moved to two-dimensional representations (e.g. iso-lines). We later used three-dimensional representations (e.g. iso-surfaces) and have arrived now at time-dependent three-dimensional representations of data that can be shown using virtual reality environments.

Two problems have arisen:

- Multi-dimensional representations of data have made it possible to show results. However, at the same time they have made it more difficult for the end user to grasp the content of what she sees in such a representation. This may be due to the complexity of the results shown. Increasingly, however, we run into problems caused by human nature. Motion sickness caused by technical devices for virtual reality representations and the tendency to filter images in order to be able to focus on what is considered to be important have become serious problems.
- The complexity of simulations makes it difficult to grasp the whole picture. Important features may not surface in multi-dimensional representations and may hence not become visible for the human spectator.

4.4 Programmability

Last but not least we are faced with the problem of programming large scale systems. Modern architectures are made of thousands of processors which have to be programmed. Current models like OpenMP and MPI have shown to be not very effective on such large scale systems.

OpenMP was designed to support shared memory programming. It was designed as an easy to program set of directives. The architecture kept in mind by the designers was a small number of processors sharing a large common memory. An extension to thousands of processes was not foreseen and is extremely difficult. Users can not expect to be able to use OpenMP as a standard programming model for large scale systems.

MPI was designed in an era when massively parallel computers were made of at most a few thousand processors and the typical systems had not more than 512 CPUs. It furthermore took into consideration the technical bound-ary conditions of the 90s with latencies of internal networks so low compared to processor speed that a software stack overhead was negligible. Today the situation has changed. Hardware latencies have decreased and the stack over-head of MPI is increasingly considered an obstacle when aiming to achieve performance. Some researchers today even consider stripping MPI off most of its functionality in order to reduce this overhead. But the MPI problem is worse. Only a small group of programmers is familiar enough with MPI to be able to write programs for 100.000 processors that not only are correct but also deliver performance. It remains to be seen whether correctness or performance is the worse problem.

5 Steps forward

Modern supercomputing architectures provide us with an unprecedented level of performance and – which may be more important – with an amount of main memory that allows us to tackle more realistic problems in more detail. At the same time these architectures introduce a number of problems that shake the foundation of our existing model of abstraction layers used in simulation. The most critical ones and some potential solutions are described here.

Simulation has to be turned into an activity that can be compared to clas-sical experiments. Ways have to be found for simulations to fulfil the require-ments of experiments. This includes standards that describe how and under which boundary conditions such simulation experiments have to be performed in order to be accepted by the scientific community. Reproducibility of such simulation experiments has to be one of the key features. The problem we have to tackle will be to define parameters of a computational environment that when kept constant always lead to the same results for simulation ex-periments. Although benchmarks like Linpack [2] or HPC Challenge [3] are a good basis to start they are mainly designed to evaluate performance. An approach to tackle reproducibility is still missing but, given the proliferation of simulation experiments, is getting increasingly important.

With this increased usage of simulation in everyday processes of research and development comes a need for verification. We need to find ways to verify that a given simulation – which is always a result of the chain of abstraction shown in figure 1 – was done correctly. This implies to define what we mean by correct and ways to prove correctness. It also implies that we need to find ways to do a verification of each step in the chain of abstraction described above. Only if we are able to solve this problem we will be able to introduce simulation experiments as a valuable resource of information and decision making tool in everyday life.

If we succeed to solve these problems we can safely assume that our results are correct – in a sense described above and we are faced with the challenge to make these results visible for experts and non-experts alike. With an ever growing spreading of simulation the focus will have to change to non-experts. For them visualization will be the method of choice to understand simulation results. Brut force virtual and augmented reality will certainly make an impression on such untrained people. However, with these complex techniques comes the problem of misinterpretation of results. As human beings we are eager to simplify and filter visual input. While trained engineers and scientists do this keeping the technical and scientific background in mind, lay people focus on the graphically important features. These may not be the technically most important ones. We hence need to find a way to deal with psychological aspects on the one hand and to automatically detect and highlight important features. Automatic pattern recognition will become an important feature but will require scientists and engineers capable of describing complex patterns. At the same time we will have to deal with the physiological side effects of modern visualization methods like motion sickness. This is a field where we may be able to learn from flight simulators and similar tools.

Finally we need to make sure that the increasing complexity of supercomputer architectures can be mastered by a sufficiently large group of experts in order to create the necessary software base to exploit their performance. It is currently not evident which methods this might be. Traditional programming languages like FORTRAN emphasize performance programming and are still close enough to the processor architecture to deliver on their promise. However, they show a lack of adaptability to complex problems that go beyond the handling of vectors and matrices. Modern languages like Java are easier to understand for the average user and complex methods can easily be expressed. But they tend to ignore the requirements of computers. For parallel programming models the situation is worse. None of the existing widely used parallel programming models (OpenMP, MPI) seems to be able to exploit large scale systems for a large group of applications. New concepts are seen in research but only a few have reached a level of maturity that seems feasible for usage in production. Among them are Co-Array FORTRAN [15] and the Unified Parallel C Project [16].

6 Conclusion

The chain of abstraction in supercomputing is based on a long history of evolution. Simulation experiments following that concept have reached a level of maturity that allows usage of simulations as standard tools. New architectures, however, have introduced a challenge to this concept. First of all through their shear level of performance and size of memory, that has pushed simulation into the production arena within only a few years. In order to fully assume its role as a tool equivalent to experiments simulation has to resolve the issues of reliability, verification and reproducibility. On the other hand modern architectures introduce a new level of complexity at the level of hardware, software, and understanding. The community will have to develop tools and methods to master this complexity. A simple extrapolation from existing moderate parallel methods will not be enough.

References

1. Nagel WE, Jäger W, Resch M (eds.) (2005) High Performance Computing in Science and Engineering 05. Springer, Berlin Heidelberg New York
2. www.top500.org
3. http://icl.cs.utk.edu/hpcc/
4. OpenMP Standard Definition www.openmp.org/
5. (1995) MPI Forum, MPI: A Message-Passing Interface Standard, Document for a Standard Message-Passing Interface, University of Tennessee
6. (1997) MPI Forum, MPI2: Extensions to the Message-Passing Interface Standard, Document for a Standard Message-Passing Interface, University of Tennessee
7. Keller R, Gabriel E, Krammer B, Müller MS, Resch MM (2003) J Grid Computing 1(2):133–149
8. Andrews PL, Banister B, Kovatch P (2004) High Performance Grid Computing via Distributed Data Access. In: Proc. of the 2004 International Conf. on Parallel and Distributed Processing Techniques and Applications, Las Vegas, NV, June 2004
9. Resch M (2004) Comp Technol 9(6):34–39
10. Ramanathan RM (2005) Intel multi-core processors: leading the next digital revolution, Intel Whitepaper ftp://download.intel.com/technology/computing/multi-core/multi-core-revolution.pdf
11. (2005) Multi-core processors – the next evoluiton in computing, AMD, Whitepaper http://multicore.amd.com/WhitePapers/Multi-Core_Processors_WhitePaper.pdf
12. Haff G (2005) Blue Gene's teraflop attack, Whitepaper, Illuminata Inc. www-03.ibm.com/servers/deepcomputing/pdf/teraflopattackilluminata.pdf
13. www.teraflop-workbench.org
14. NPARC Alliance CFD Verification and Validation Web Site http://www.grc.nasa.gov/WWW/wind/valid/tutorial/bibliog.html
15. http://www.co-array.org/
16. http://upc.lbl.gov/

3D Euler flow simulation in hydro turbines: unsteady analysis and automatic design

S. Cherny[1], D. Chirkov[1], V. Lapin[1], I. Lobareva[1], S. Sharov[1], and V. Skorospelov[2]

[1] Institute of Computational Technologies SB RAS, Lavrentiev Ave. 6, 630090 Novosibirsk, Russia cher@ict.nsc.ru
[2] Sobolev Institute of Mathematics SB RAS, Lavrentiev Ave. 4, 630090 Novosibirsk, Russia vskrsp@math.nsc.ru

Summary. Two hydro turbine fluid flow problems are solved numerically in the paper. Both problems are solved in frames of Euler model using CFD code developed by the authors. The code is based on finite volume artificial compressibility approach. The accuracy of numerical scheme is second order in time and third order in space. For the solution of unsteady problems dual-time-stepping algorithm is used. Complex geometry of a turbine passage is handled using domain decomposition. The first problem considered is a simulation of precessing vortex rope downstream the turbine runner. Numerical results showed that this phenomenon can be captured by Euler model. The second problem is an automatic design (optimization) of runner blade shape with help of Breeder genetic algorithm. Different formulations of the objective function are considered and their influence on the blade shape is demonstrated.

1 Introduction

Fluid flow simulation in hydro turbine components is an urgent and interesting CFD problem. Main features of this problem are complex 3D geometry of the turbine, flow rotation and presence of turbulence, especially in the draft tube. However in some situations, concerning spiral case, distributor, and even runner simulations the effect of viscosity can be neglected, and inviscid incompressible model can be used to calculate the flow-field. Present work is focused on two of such problems, solved using artificial compressibility finite volume code, developed by the authors for the solution of Euler equations. Presented results continue previous investigations published in works [1] — [10].

This work was supported by the Russian Foundation for Basic Research (project No. 04-01-00246) and the Integration Basic Research Program No. 27 of the SB RAS.

Cone diffuser of hydro turbine draft tube, situated right downstream the runner, is the place of pronounced flow non-stationarity. Non-stationary phenomena become apparent in off-design operating regimes as a curved vortex rope rotating around the axis of the cone. Precessing vortex rope has a strong influence on the flow-field upstream, as well as downstream, resulting in dynamic forces acting on the blades and draft tube walls. It is considered, that fluid viscosity is the main factor responsible for the formation of precessing vortex rope. According to this it's adequate simulation requires a sophisticated turbulent model, capable to resolve vortices comparable to the diameter of the cone. To prove this, in [11, 9] calculations of turbulent flow in a straight cone diffuser has been performed. Velocity distribution at the inlet was steady-state and axially symmetric, but similar to that, observed behind the runner at part load regimes. VLES [11] and LES [9] simulations produced precessing vortex rope, while RNG $k - \varepsilon$ model gave steady-state axially symmetric flow. However we suppose that there are also some different factors responsible for formation of precessing vortex rope, such as circumferential non-uniformity of flow-field upstream. In order to analyze this hypothesis unsteady Euler simulations of full turbine passage has been carried out in present work.

Using advanced computational methods and modern high-speed computers makes it possible to automate runner design process. Ideally the automatic design is seen as a sequential simulation of different runner with perturbed configurations in order to choose the best one from point of view of some a priori specified criterion, for example, of turbine efficiency. To cope this one have for each runner geometry to adequately simulate turbulent flow in the whole turbine including spiral case, distributor, runner and draft tube. Such calculations are possible but computationally expensive. A simplified approach has been used in works [12, 13], where runner optimization was done by means of $k - \varepsilon$ turbulent flow simulations only in the runner.

At the same time 3D Euler calculation of flow inside one runner blade channel takes only a few minutes. This fact and the observation, that influence of turbulence and viscosity in the runner is small enough, lead us to the idea of runner optimization in frames of Euler equations. Of coarse, this statement raises a problem of appropriate objective functional choice. Some approaches to manage it are discussed in [14].

Automatic runner blade optimization system based on Euler flow-field computations is described in the second part of the paper. Discussed are the issues concerning geometry parameterization, objective functional choice, and strategy of global extremum search. The capabilities of the system are demonstrated in several optimization runs for Bratskaya hydro power station refurbishment project.

2 Problem statement

2.1 Governing equations

For the results, presented in this paper a 3D unsteady, inviscid and incompressible flow throughout a turbine is considered. Flow motion in static elements (spiral case, distributor and draft tube) is described in absolute reference frame, and rotating reference frame y_1, y_2, y_3 is used for the runner, rotating with constant angular velocity ω. Absolute velocity vector $\mathbf{v}$ and relative velocity vector $\mathbf{w} = (w_1, w_2, w_3)$ are connected by the following relation

$$\mathbf{v} = \mathbf{w} + \boldsymbol{\omega} \times \mathbf{r}, \tag{1}$$

where $\boldsymbol{\omega} = (0, 0, \omega)$ is an angular velocity vector, and $\mathbf{r}$ is a radius-vector of the point.

Euler equations for the relative motion are:

$$\frac{\partial w_j}{\partial y_j} = 0, \tag{2}$$

$$\frac{\partial}{\partial t} w_i + \frac{\partial}{\partial y_j}(w_i w_j) + \frac{\partial}{\partial y_i} p = f_i, \tag{3}$$

where $\mathbf{f} = (f_1, f_2, f_3) = -\boldsymbol{\omega} \times \boldsymbol{\omega} \times \mathbf{r} - 2\boldsymbol{\omega} \times \mathbf{w}$. Vector $\boldsymbol{\omega} \times \boldsymbol{\omega} \times \mathbf{r}$ is the centripetal acceleration, $2\boldsymbol{\omega} \times \mathbf{w}$ is the Coriolis acceleration.

2.2 Boundary conditions

For the results presented in this paper two problem statements were considered.

1. Unsteady flow simulation in the whole turbine. In this case computational domain included 60 blocks, covering spiral case, all distributor channels, all runner channels, and draft tube cone (fig. 1), following idea of domain decomposition. At the entrance of spiral case uniform velocity distribution had been specified corresponding to a given discharge Q. Internal boundaries are treated by exchanging fluid flow parameters between neighbouring blocks on each iteration.

Common approach for outlet boundary is to set fixed pressure distribution. However, when outlet boundary is situated in the draft tube cone, more accurate results can be obtained using radial equilibrium condition for pressure. This condition can be derived from radial component of momentum equation

$$c_r \frac{\partial c_r}{\partial r} + \frac{c_u}{r}\frac{\partial c_r}{\partial \theta} + c_z \frac{\partial c_r}{\partial z} - \frac{c_u^2}{r} = -\frac{\partial p}{\partial r} \tag{4}$$

assuming that radial velocity component c_r is small enough and flow is axially symmetric. In this case equation (4) at $z = const$, corresponding to outlet section of the cone, turns into ordinary differential equation

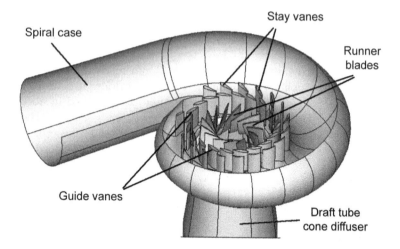

Fig. 1. Computational domain for unsteady flow simulation in the whole turbine passage

$$\frac{dp}{dr} = \frac{c_u^2}{r}. \tag{5}$$

Pressure distribution in the outlet section is obtained through integration of this equation along the radius from the draft tube wall (where constant pressure p_{out} is specified) to rotation axis.

2. Steady periodic flow simulation in distributor and runner. Here calculations are performed only in one guide vane channel of distributor and one runner blade channel assuming that flow-field in the rest channels are the same (fig. 2). It was achieved using periodic boundary condition. In order to transmit information from static distributor to rotating runner and vise versa, flow parameters were averaged in circumferential direction.

Radial equilibrium condition for pressure is used in outlet section.

3 Numerical method

3.1 Artificial compressibility finite volume method

Governing equations are solved numerically using artificial compressibility approach and implicit finite volume method [1, 2]. Following the idea of artificial compressibility, pseudotime derivative of pressure is added to continuity equation (2):

$$\frac{\partial p}{\partial \tau} + \beta \frac{\partial w_j}{\partial y_j} = 0, \tag{6}$$

where β is an artificial compressibility coefficient. Also pseudotime derivatives of velocity components are added to momentum equations (3)

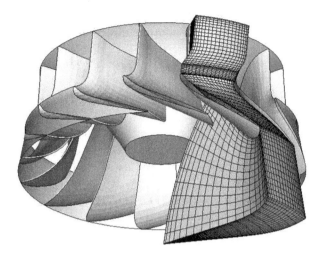

Fig. 2. Computational domain and mesh for periodic runner flow simulation

$$\left(\frac{\partial}{\partial \tau} + \frac{\partial}{\partial t}\right) w_i + \frac{\partial}{\partial y_j}\left(w_i w_j\right) + \frac{\partial}{\partial y_i} p = f_i. \tag{7}$$

To construct conservative numerical scheme, modified equations (6), (7) are presented in form of conservation laws for arbitrary fixed volume V

$$(\mathbf{R}^\tau \frac{\partial}{\partial \tau} + \mathbf{R}^t \frac{\partial}{\partial t}) \int_V \mathbf{Q} dV + \oint_{\partial V} \mathbf{H} \cdot d\mathbf{S} = \int_V \mathbf{F} dV, \tag{8}$$

where ∂V is the surface of volume V; $d\mathbf{S} = \mathbf{n} \cdot dS$ is the surface element, multiplied by unit normal vector; $\mathbf{Q} = (p, w_1, w_2, w_3)^T$; $\mathbf{H} = (\mathbf{w}, w_1 \mathbf{w} + p\mathbf{e_1}, w_2 \mathbf{w} + p\mathbf{e_2}, w_3 \mathbf{w} + p\mathbf{e_3})^T$; $\mathbf{F} = (0, f_1, f_2, f_3)^T$; $\mathbf{R}^\tau = diag(1, 1, 1, 1)$; $\mathbf{R}^t = diag(0, 1, 1, 1)$; $\mathbf{e_1} = (1, 0, 0), \mathbf{e_2} = (0, 1, 0), \mathbf{e_3} = (0, 0, 1)$ is a basis of Cartesian coordinate system y_1, y_2, y_3.

Let us introduce the following notation for $\mathbf{Q}$ and $\mathbf{F}$, averaged over cell $i\,j\,k$ with volume V_{ijk}

$$\mathbf{Q}_{ijk}^n = \frac{1}{V_{ijk}} \int_{V_{ijk}} \mathbf{Q}^n dV, \qquad \mathbf{F}_{ijk}^n = \frac{1}{V_{ijk}} \int_{V_{ijk}} \mathbf{F}^n dV$$

and attribute them to the cell center. Here superscript n is the number of time layer.

Discretization of (8) gives:

$$\left[\mathbf{R}^\tau \frac{\left(\mathbf{Q}^{n+1}\right)^{s+1} - \left(\mathbf{Q}^{n+1}\right)^s}{\Delta\tau} + \mathbf{R}^t \frac{3\left(\mathbf{Q}^{n+1}\right)^{s+1} - 4\mathbf{Q}^n + \mathbf{Q}^{n-1}}{2\Delta t}\right] V = \left(\mathbf{RHS}^{n+1}\right)^{s+1}, \tag{9}$$

where $\Delta\tau$ and Δt are the pseudotime step size and time step size, respectively. s is the number of iteration in pseudotime direction. Right hand side is

$$\mathbf{RHS} = -\left(\mathbf{H}_{i+1/2} - \mathbf{H}_{i-1/2} + \mathbf{H}_{j+1/2} - \mathbf{H}_{j-1/2} + \mathbf{H}_{k+1/2} - \mathbf{H}_{k-1/2}\right) + \mathbf{FV},$$

$$\mathbf{H}_{m+1/2} = \frac{1}{2}\left[(\mathbf{H}_{m+1} + \mathbf{H}_m)\cdot\mathbf{S}_{m+1/2} - |\mathbf{A}|_{m+1/2}\,\Delta_{m+1/2}\mathbf{Q}\right] - \mathbf{W}_{m+1/2},$$

$$\mathbf{A} = \frac{\partial(\mathbf{H}\cdot\mathbf{S})}{\partial\mathbf{Q}} = \mathbf{A}^+ + \mathbf{A}^-, |\mathbf{A}| = \mathbf{A}^+ - \mathbf{A}^-.$$

Flux vector jacobian $\mathbf{A}$ can be presented as $\mathbf{A} = \mathbf{RDR}^{-1}$, where $\mathbf{R}$ is the right eigenvector matrix; $\mathbf{D} = diag\,(\lambda_1, \lambda_2, \lambda_3, \lambda_4)$, $\lambda_{1,2} = U$, $\lambda_{3,4} = U \pm \sqrt{U^2 + \beta\mathbf{S}\cdot\mathbf{S}}$, $U = \mathbf{w}\cdot\mathbf{S}$, $\mathbf{A}^\pm = 0.5\mathbf{R}\,(\mathbf{D} \pm |\mathbf{D}|)\,\mathbf{R}^{-1}$. $\mathbf{W}$ consists of additional terms, to give a resulting scheme of 3-rd order spatial accuracy.

If pseudotime iterations converged, (9) becomes a system of discrete equations, approximating integral form of Euler equations (2), (3) with second order in time and third order in space.

In order to provide fast convergence and reduce numerical dissipation of the resulting scheme artificial compressibility coefficient β in the present work was taken

$$\beta = \delta W_{ref}^2,$$

where $\delta = 5 \div 10$, W_{ref} is a magnitude of characteristic speed of the problem.

3.2 Solution of nonlinear equations

Equations (9) are linearized using Newton method

$$\left[\left(\frac{1}{\Delta\tau}\mathbf{R}^\tau + \frac{3}{2\Delta t}\mathbf{R}^t\right)V - \left(\frac{\partial}{\partial\mathbf{Q}}\mathbf{RHS}\right)^s\right](\mathbf{Q}^{s+1} - \mathbf{Q}^s) =$$
$$-\mathbf{R}^t\frac{3\mathbf{Q}^s - 4\mathbf{Q}^n + \mathbf{Q}^{n-1}}{2\Delta t} + \mathbf{RHS}^s, \qquad (10)$$

here time superscript $n+1$ is omitted.

Leaving only first order differences in evaluation of term $\partial\mathbf{RHS}/\partial\mathbf{Q}$, we obtain from (10) the following system of linear equations

$$\left[\left(\frac{1}{\Delta\tau}\mathbf{R}^\tau + \frac{3}{2\Delta t}\mathbf{R}^t\right)V + \mathbf{A}_{i+1/2}^-\Delta_{i+1/2} + \mathbf{A}_{i-1/2}^+\Delta_{i-1/2} + \right.$$
$$\mathbf{A}_{j+1/2}^-\Delta_{j+1/2} + \mathbf{A}_{j-1/2}^+\Delta_{j-1/2} +$$
$$\left.\mathbf{A}_{k+1/2}^-\Delta_{k+1/2} + \mathbf{A}_{k-1/2}^+\Delta_{k-1/2}\right]\Delta^{s+1}\mathbf{Q} =$$
$$-\mathbf{R}^t\frac{3\mathbf{Q}^s - 4\mathbf{Q}^n + \mathbf{Q}^{n-1}}{2\Delta t}V + \mathbf{RHS}^s, \qquad (11)$$

where $\Delta^{s+1}\mathbf{Q} = \mathbf{Q}^{s+1} - \mathbf{Q}^s$.

After approximate LU-factorization, system (11) is resolved in 2 sequential steps:

$$\Delta^{s+1/2}\mathbf{Q}_{ijk} = \mathbf{B}^{-1}\left[-\mathbf{R}^t\frac{1}{2\Delta t}(3\mathbf{Q}^s - 4\mathbf{Q}^n + \mathbf{Q}^{n-1}) + \mathbf{RHS}^s +\right.$$

$$\left.\mathbf{A}^+_{i-1/2}\Delta^{s+1/2}\mathbf{Q}_{i-1jk} + \mathbf{A}^+_{j-1/2}\Delta^{s+1/2}\mathbf{Q}_{ij-1k} + \mathbf{A}^+_{k-1/2}\Delta^{s+1/2}\mathbf{Q}_{ijk-1}\right],$$

$$\Delta^{s+1}\mathbf{Q}_{ijk} = \Delta^{s+1/2}\mathbf{Q}_{ijk} - \mathbf{B}^{-1}\left(\mathbf{A}^-_{i+1/2}\Delta^{s+1}\mathbf{Q}_{i+1jk} +\right.$$

$$\left.\mathbf{A}^-_{j+1/2}\Delta^{s+1}\mathbf{Q}_{ij+1k} + \mathbf{A}^-_{k+1/2}\Delta^{s+1}\mathbf{Q}_{ijk+1}\right),$$

where

$$\mathbf{B} = \left(\frac{1}{\Delta\tau}\mathbf{R}^\tau + \frac{3}{2\Delta t}\mathbf{R}^t\right)V +$$

$$\mathbf{A}^+_{i-1/2} - \mathbf{A}^-_{i+1/2} + \mathbf{A}^+_{j-1/2} - \mathbf{A}^-_{j+1/2} + \mathbf{A}^+_{k-1/2} - \mathbf{A}^-_{k+1/2}.$$

4 Unsteady simulation of vortex rope in the draft tube

This section presents the results of unsteady simulation of Francis turbine with RO910 runner, set at Platanovryssi hydro power station (Greece). Reduced parameters of operating regime are: runner diameter $D'_1 = 1$m, head $H' = 1$m, runner frequency $n'_1 = 73.5$rpm, gravity acceleration $g' = D_1 g/H = 0.188$. Time step $\Delta t = 0.0085$ s corresponded to the rotation of runner at angle $\Delta\phi = 3.75°$. 6 time steps needed for the runner to sweep one blade channel, and 96 time steps to rotate at 360°. Runner frequency is 1.225 Hz, while the frequency, at which blades pass the fixed static point is 19.6 Hz. Two operating regimes are considered.

4.1 Regime of nominal power

This regime corresponds to reduced discharge $Q'_1 = 1.004$ m^3/s, set by wide opening of distributor guide vanes. After the simulation, pressure and velocity field were analyzed behind the runner. Axially symmetric rotating flow with negligible precession has been observed. The absence of precession was also indicated in experiment on a test rig.

Fig. 3 compares circumferential c_u and axial c_z components of absolute velocity along the radius behind the runner, obtained experimentally (■) and numerically in full turbine computation (△) and periodic computation of only one runner blade channel (◇). It is evident, that full turbine computation gives more accurate results.

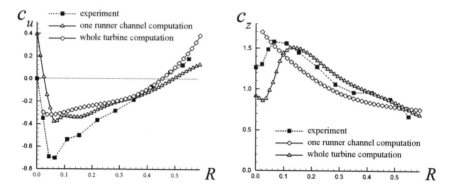

Fig. 3. Comparison of swirl c_u and axial c_z velocity components at nominal power regime

4.2 Part load regime

Part load regime corresponds to reduced discharge $Q'_1 = 0.5975$ m^3/s, and narrow opening of distributor guide vanes.

Instant flow pictures for computation and experiment behind the runner are shown in Fig. 4. In case of numerical result vortex rope was visualized by iso-pressure surface. In experiment visualization of vortex rope was done by injecting air bubbles into the flow, and drawing the place of their accumulation. In computation, as well as in the experiment pronounced vortex rope precession is observed. Fig. 5 shows pressure distribution in horizontal cross-section of draft tube cone in different moments of time, indicating repeated behaviour of the vortex core with period 2.652 seconds (0.38 Hz). The ratio of runner frequency to precession frequency is 3.25. Note that different experimental data give the range $2 \div 5$ for this ratio.

Fig. 6 illustrate pressure fluctuations in point "1", shown in figure 4, and its Fourier transformation. High peak on the frequency-amplitude graph corresponds to vortex precession frequency (0.38 Hz). Small peak corresponds to frequency 19.6 Hz of pressure pulsations, caused by runner blades passing. It is evident, that vortex rope pulsations dominate in point "1".

In order to investigate the upstream influence of vortex precession, pressure fluctuations in static points "2" and "3" (see Fig. 4), were also analyzed. Results are plotted in Fig. 7 and 8. Two main peaks in each Fourier transformation graph indicate main frequencies and amplitudes of fluctuations. It can be seen that pressure fluctuations in each point contain a harmonic, corresponding to vortex rope precession. Thus vortex precession affects the flow-field in the runner and even upstream the runner. However this influence decreases with moving off the vortex location. At the same time the amplitude of harmonic, having blade passing frequency (19.6 Hz) increases, reaching maximum in point "3" before the inlet edges of the blades.

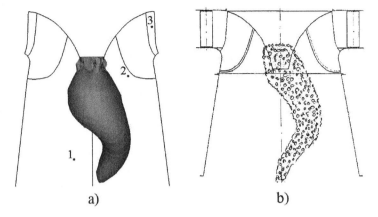

Fig. 4. Draft tube vortex rope in part load regime: a) - computation; b) - experiment

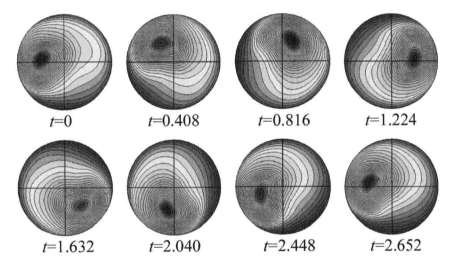

Fig. 5. Pressure distribution in different time steps in draft tube cross section

In present investigation, in contrast to [11, 9], flow-field non-uniformity at the inlet of cone diffuser is taken into account. Also computations were carried out in frames of Euler equations, so physical viscosity is absent in the model. However numerical simulation in this statement also give a precessing vortex rope, being in good agreement with experiment in terms of vortex shape and precession frequency. We conclude that vortex formation can be caused by upstream flow-field non-uniformity and artificial viscosity, introduced by numerical scheme dissipation.

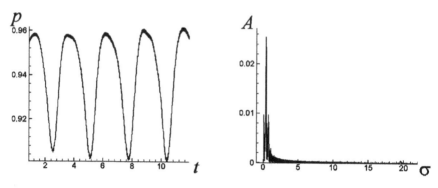

Fig. 6. Pressure fluctuations in point "1" (draft tube) and its Fourier transformation

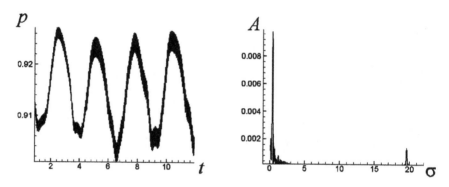

Fig. 7. Pressure fluctuations in point "2" (after the blade) and its Fourier transformation

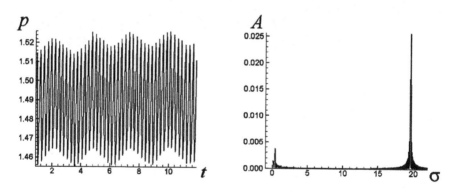

Fig. 8. Pressure fluctuations in point "3" (before the blade) and its Fourier transformation

5 Runner blade shape optimization

5.1 Blade parameterization

Runner blade variation is done by variation the values of parameters defining the blade shape. Blade parameterization should be flexible enough, but on the other hand, the number of free parameters should be as small as possible. In the present work we use so called "relative" parameterization of blade median surface, containing 16 free parameters.

A Cartesian coordinate system $OXYZ$ is assumed to be assigned with the runner, with OZ axis coinciding with the runner rotation axis and directed downwards. Runner blade geometry can be represented as a combination of its median surface

$$\mathbf{r}(u,v) = \{X(u,v), Y(u,v), Z(u,v)\}, \qquad u, v \in [0,1]. \tag{12}$$

and thickness distribution function $d = d(u,v)$. Here (u,v) — is a "natural" parameterization of blade median surface (fig. 9). Median surface equation (12) can be written in cylindrical coordinate system:

$$R = R(u,v) = \sqrt{X(u,v)^2 + Y(u,v)^2}, \qquad Z = Z(u,v),$$

$$\varPhi = \varPhi(u,v) = \operatorname{arctg}\frac{Y(u,v)}{X(u,v)}$$

The shape of median surface is determined mainly by angular function $\varPhi(u,v)$, therefore blade variation during the optimization is done by means of varying only this function $\varPhi(u,v)$, leaving RZ-projection and initial thickness distribution $d(u,v)$ unchanged.

Let $\varPhi_0(u,v)$ be the angular function of initial blade, and $\varPhi(u,v)$ the angular function of perturbed blade. For blade shape variation we use "relative" approach. Function, subjected to parameterization is not $\varPhi(u,v)$, but its deviation $\varPhi'(u,v) = \varPhi(u,v) - \varPhi_0(u,v)$ from the initial one. As an optimization space we consider the set of $\varPhi'(u,v)$ functions which are bicubic polynomials. Finally, variation of $\varPhi'$ is done by means of variation of 16 coefficients φ_{ij} of bicubic polynomial, written in Bernstein's form:

$$\varPhi'(u,v) = \sum_{i=0}^{3}\sum_{j=0}^{3} \varphi_{ij} B_i(u) B_j(v), \tag{13}$$

where $B_k(w) = \dfrac{3!}{k!(3-k)!} w^k (1-w)^{3-k}$. The main advantage of such "relative" parameterization is that initial blade, which have a complex form and generally is not a bicubic polynomial, always belongs to optimization space (when $\varphi_{ij} = 0$).

5.2 3D Euler flow-field calculations

For each perturbed blade shape 3D steady-state Euler flow-field calculation is performed using algorithm, described above.

Flow calculations were performed only in one runner blade channel assuming that flow-field in the rest channels are the same (fig. 2). Besides the computational domain geometry, the distributions of a velocity vector at inlet cross-section AA' and pressure distribution in the outlet cross-section BB' (fig. 10), as well as rotation frequency of the runner should be set as input data for the calculation. These parameters are assumed to be constant during the process of optimization and correspond to the predicted best efficiency operating point for initial blade geometry. A steady-state field achieved at the $(i-1)$-th optimization step was taken as an initial guess for the i-th step flow-field calculation, $i = 2, 3, \ldots$. This approach allowed us to save about 50% of CPU time.

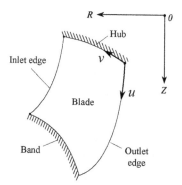

Fig. 9. RZ-projection of the blade

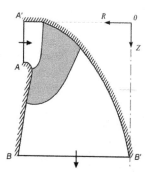

Fig. 10. Meridional projection of computational domain

5.3 Constraints

Besides linear constraints for geometry parameters, runner internal head and cavitation constraints were also set for the optimization problem.

Internal head constraint. In order optimal runner correspond to a given turbine operation regime it is necessary to ensure that change in total energy between inlet section of spiral case and outlet section of the draft tube of turbine coincides with the given net head H of the plant. However during the optimization we compute the flow-field only in the runner. Therefore, in order to maintain given net head H, the calculated runner internal head H_c should be close to it's target value, which for Bratskaya hydro power station can be estimated as $0.95H$. The resulting internal head constraint had the form:

$$|H_c - 0.95H| \leq \varepsilon, \tag{14}$$

where ε is a small parameter.

Cavitation constraint. One of the main characteristics of turbine operation is its cavitation characteristic, determining the presence and size of cavitation regions. Cavitation arises when static pressure in a liquid flow becomes lower than the vapor pressure at a given temperature:

$$p \leq p_v. \tag{15}$$

In runner flow simulations based on Euler equations without the draft tube pressure flow-field is determined up to a constant p_0, set at the outlet section of computational domain. In order to obtain absolute value of the pressure one have to estimate absolute value of p_0. It can be estimated using Bernoulli equation for the streamline going from draft tube inlet to draft tube outlet.

It is well known that suction sides of the blades are mostly subjected to cavitation. In case the area of cavitation region exceeds the threshold value of approximately 15 percent of suction side area the breakdown in efficiency and power output, as well as strong vibration are observed. It is not always possible to completely prevent cavitation in the runner blade channel. In many cases it is reasonable to design the runner so that the area of cavitation region doesn't exceed the value of $0.15S_{suc}$, where S_{suc} is the area of suction side of the blade. That's why the cavitation restriction had the form:

$$\frac{S_{cav}}{S_{suc}} \leq 0.15, \tag{16}$$

where S_{cav} is the area on suction side, where (15) is fulfilled.

5.4 Objective functionals

Ideally shape optimization applied to hydro turbines comes to efficiency maximization at given operation regime, with some constraints imposed on cavitation and structural strength. In case of Euler calculations only in the runner, explicit evaluation of losses is impossible, therefore one have to formulate objective functional, that would indirectly estimate efficiency. Presented below are some particular formulations of objective functionals, based on the experience of runner designers.

1. Runner outlet kinetic energy. Assume that the whole kinetic energy leaving the runner is lost (which is not true in reality). Then most efficient operation of the turbine correspond to the minimal kinetic energy at runner outlet. We considered two energy objective functionals, corresponding to two different cross sections behind the runner:

$$E_1 = \frac{1}{Q} \int_{S_1} \frac{\mathbf{v}^2}{2g} (\mathbf{v} \cdot d\mathbf{S}), \tag{17}$$

$$E_2 = \frac{1}{Q} \int_{S_2} \frac{\mathbf{v}^2}{2g} (\mathbf{v} \cdot d\mathbf{S}). \tag{18}$$

S_1 is the cross section just after the blades, whereas S_2 is situated at the inlet to the draft tube. $\mathbf{v}$ — is the absolute velocity vector; $d\mathbf{S}$ — vector, normal to surface element dS, of length equal to area dS. It can be shown that minimum of E_2 correspond to a homogeneous distribution of axial velocity component c_z with zero swirl component in the outlet cross section. Minimization of E_1 is directed to reduce inductive losses in the runner, whereas minimization of E_2 should enhance draft tube operation.

2. $\mathbf{F_{cav}}$ — relative size of cavitation zone. Objective function is a relative size of cavitation zone on suction size

$$F_{cav} = \frac{S_{cav}}{S_{suc}}. \tag{19}$$

3. $\mathbf{F_S}$ — deviation of blade surface streamlines from average stream direction. Corresponding objective functional can be formulated as follows:

$$F_S = \frac{1}{S} \int_S (1 - \sigma(\beta) \cos \beta) \, dS, \tag{20}$$

where S is the area of blade surface, β is the angle between surface streamline and corresponding average stream direction (fig. 11), and weight function σ has the form

$$\sigma(\beta) = \begin{cases} 1, & \beta < \pi/2, \\ \sigma_0, & \beta \geq \pi/2 \end{cases}$$

Parameter $\sigma_0 \geq 1$ has been introduced in order to suppress backflows, substantially increasing the value of functional. With appropriate value of σ_0 functional (20) allows us to fulfill one of the main requirements to runner blade shape: zero angle of attack (flow diverging line at the leading edge coincides with the blade tip).

5.5 Optimization method

Mathematically optimization problem is formulated as:
    find

$$\min F(\mathbf{x}), \qquad \mathbf{x} = (x_1, \ldots, x_N) \in X$$

with constraints:

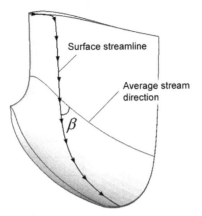

Fig. 11. Surface streamlines on pressure side

$$E_N \supset X = \{\mathbf{x} : x_i' \le x_i \le x_i''\},$$
$$\psi_k(\mathbf{x}) \le 0, \quad k = 1, \dots, m.$$

In order to take into account the imposed constraints the compound functional with penalty functions is used:

$$F_c = F \left[1 + \sum_{j=1}^{m} K_j(\psi_j + |\psi_j|) \right].$$

Searching of global minimum of compound objective function is done using Breeder Genetic Algorithm (BGA) [14]. First, a set of p randomly perturbed runners (generation of individuals) is formed. Each individual is the runner, described by the vector of geometry parameters $\mathbf{x} = (x_1, \dots, x_N)$. For each runner flow-field is calculated and the value of objective functional is evaluated. Then the best 30% of runners are selected and using them as parents, new generation is constructed, following processes of recombination, mutation and cloning. Then loop repeats(fig. 12). For the results presented below number of individuals in generation is $p = 70$ and number of generations calculated is $N_{Gen} = 30$.

5.6 Results of optimization calculations

Francis runner of Bratskaya hydro power plant has been taken as an object of optimization Fig 2 shows the runner and computational domain with mesh having $40 \times 20 \times 20$ cells. Steady-state solution with $\text{div}(\mathbf{v}) \le 10^{-3}$ on this mesh takes from 1000 to 4000 time steps, or 1 to 4 minutes of Athlon 2500+ CPU time. Creation of 30 generations with 70 individuals in each generation result in 2100 runner flow-field computations, taking about 3 days.

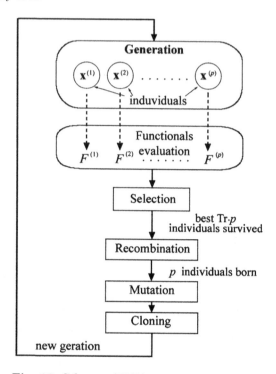

Fig. 12. Scheme of BGA optimization process

Ideally one would wish to minimize all the presented functionals. However this task appeals to multiobjective optimization, which is beyond the scope of this paper. At the present stage it was interested for us to understand the influence of each particular functional minimization on the blade shape.

Outlet kinetic energy minimization. Fig. 13 presents distributions of meridional velocity for the initial runner and optimal runners obtained through minimization of E_1 (17) and E_2 (18) energies. Runner head constraint (14) and cavitation constraint (16) were imposed in optimization. As it was expected, in case of E_1 minimization velocity distribution in section S_1 right behind the blade become more perpendicular to the cross section. For runner with optimal E_2 energy in velocity distribution in outlet section is much more uniform, with 1.5 times less kinetic energy compared to initial one. Fig. 13 also indicates that kinetic energies E_1 and E_2 are rather independent characteristics: minimization of E_1 can increase E_2, and vice versa. It should also be noted that in both computations flow diverging line at inlet edge moved to pressure side, which is undesirable.

Cavitation minimization. Fig. 14 shows suction side pressure distributions for initial and optimized blade, obtained with the only constraint (14). Shaded area is the cavitation zone, where condition (15) is fulfilled. As the result of optimization, cavitation zone decreased from 5% to approximately 1% of the whole suction side.

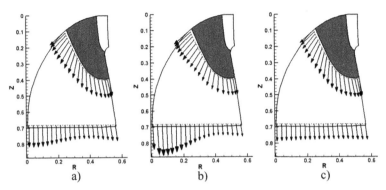

Fig. 13. Meridional velocity, averaged in circumferential direction. a) — initial runner; b) — runner with optimal E_1; c) — runner with optimal E_2

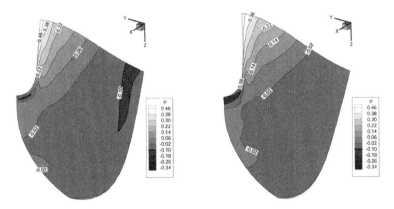

Fig. 14. Pressure distribution and cavitation zones on suction side for F_{cav} minimization. Left is initial blade, right is the optimal blade

Streamline optimization. Fig. 15 illustrates the results of streamline deviation optimization (20) with $\sigma_0 = 10$ and constraints (14) and (16). With using $\sigma_0 = 1$ it was possible to significantly straighten the flow and reduce deviation of blade streamlines from average stream direction, but in this case flow diverging line on inlet edge shifted to pressure side, indicating high angle of attack. The use of $\sigma_0 = 10$ allowed to completely remove fluid flow from pressure side to suction side at the inlet edge, ensuring zero angle of attack. Meantime general flow behaviour was not significantly affected (see Fig. 15).

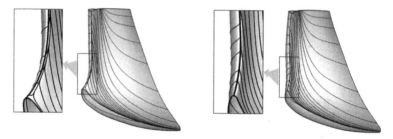

Fig. 15. Surface streamlines at inlet edge for F_S minimization. Left is initial blade, right is the blade giving minimum to F_S (20) with $\sigma_0 = 10$

5.7 Discussion

It follows form the optimization results that created tool allows us to effectively enhance the quality of the blade with respect to one fixed criterion. However, enhancing one characteristic usually deteriorates the others. Therefore we have to enhance, or at least control, several criteria simultaneously. To solve this task there are different approaches. One of them, used in [14], is to minimize the weighted sum of all functionals. It's drawback: for adequate choice of the weights for a given turbine, experienced designer have to collect and analyze the results of big number of optimization runs with different combinations of weights. The second approach suggests in process of one functional minimization to impose constraints on the values of others. However, from authors' point of view, the most promising approach is multiobjective optimization, reducing several functionals at once [15]. In this case the result of optimization is not only one shape, but the set of shapes, making Pareto front.

Then, one of the further steps is the use of more flexible runner parameterization with spline-approximation of $\Phi(u, v)$, variable RZ projection and thickness distribution of the blade. The other direction of further investigations is runner efficiency optimization using turbulent flow solver.

References

1. Cherny S, Gryazin Y, Sharov S, Shashkin P (1996) An efficient LU-TVD finite volume method for 3-D inviscid and viscous incompressible flow problems. In: Tallec P (ed) Proc. of the 3rd ECCOMAS Computational Fluid Dynamics Conf. John Wiley & Sons, Paris
2. Gryazin Y, Cherny S, Sharov S, Shashkin P (1997) Dokl Akad Nauk 353:478–483 (in Russian)
3. Cherny S, Shashkin P, Gryazin Y (1999) Comp Technol 4:74–94 (in Russian)
4. Kovenya V, Cherny S, Sharov S, Karamyshev V, Lebedev A (2001) Comput Fluids 30:903–916

5. Skorospelov V, Turuk P, Cherny S, Sharov S (2001) Numerical simulation of water flow in full hydro turbine passage. In: Shokin Yu (ed) Proc. of International Conf. on Recent Developments in Applied Mathematics and Mechanics: Theory, Experiment and Practice, Novosibirsk (in Russian)
6. Kuzminov A, Lapin V, Cherny S (2001) Comp Technol 6:73–86 (in Russian)
7. Cherny S, Sharov S, Skorospelov V, Turuk P (2003) Russ J Numer Anal Math Modelling 18:87–104
8. Skorospelov V, Turuk P, Aulchenko S, Latypov A, Nikulichev Y, Lapin V, Chirkov D, Cherny S (2004) Solution of the 3D optimization problem of the aerohydrodynamic shape of turbine components. In: Fomin V (ed) Proc. of XII-th International Conf. on the Methods of Aerophysical Research, Novosibirsk (in Russian)
9. Lapin V, Cherny S, Skorospelov V, Turuk P (2004) Problems of flow simulation in turbomachnes. In: Shokin Yu (ed) Proc. of International Conf. on Computational and Informational Technologies for Research, Engineering and Education, Almaty (in Russian)
10. Sharov S, Cherny S, Skorospelov V, Turuk P (2004) Investigation of non-stationary flows in turbomachines. In: Shokin Yu (ed) Proc. of International Conf. on Computational and Informational Technologies for Research, Engineering and Education, Almaty (in Russian)
11. Ruprecht A, Helmrich T, Aschenbrenner T, Scherer T (2000) Simulation of vortex rope in a draft tube. In: Proc. of the 20th IAHR Symp. on Hydraulic Machinery and Systems, Charlotte
12. Enomoto Y, Kurosawa S, Suzuki T (2004) Design optimization of Francis turbine runner using multi-objective genetic algorithm. In: Dahlback N, Gustavsson H, Francois M (eds) Proc. of 22nd IAHR Symp. on Hydraulic Machinery and Systems, Stockholm
13. Tomas L, Pedretti C, Chiappa T, Francois M, Stoll P (2002) Automated design of a Francis turbine runner using global optimization algorithms. In: Avellan F, Ciocan G, Kvicinsky S (eds) Proc. of the 21st IAHR Symp. on Hydraulic Machinery and Systems, Lausanne
14. Sallaberger M, Fisler M, Michaud M, Eisele K, Casey K (2000) The design of Francis turbine runners by 3D Euler simulations coupled to a breeder genetic algorithm. In: Proc. of the 20th IAHR Symp. on Hydraulic Machinery and Systems, Charlotte
15. Favaretto C, Funazaki K, Tanuma T (2003) The development of a genetic algorithm code for secondary flow injection optimization in axial turbines. In: Proc. of the International Gas Turbine Congress, Tokyo

On parallelization of one 3D fluid flow simulation code

T. Bönisch[1], G.S. Khakimzyanov[2], and N.Yu. Shokina[1,2]

[1] High Performance Computing Center Stuttgart (HLRS), University of Stuttgart, Nobelstraße 19, 70569 Stuttgart, Germany boenisch@hlrs.de, shokina@hlrs.de
[2] Institute of Computational Technologies SB RAS, Lavrentiev Ave. 6, 630090 Novosibirsk, Russia khak@ict.nsc.ru

Summary. The effectiveness of the parallelization of a 3D fluid flow simulation code using Software Engineering principles is considered. The problem on three-dimensional stationary ideal fluid flows through channels with complicated geometry has been investigated. The realization of the numerical algorithm has led to the sequential program, which has to be parallelized in order to reduce the execution time. The analysis of the sequential code is done in order to choose the parallelization approach, specify the kind of parallelization method, make the decision about the parallelization approach for the solver and define the target computer architectures. The results of parallel computations are provided for several hardware platforms.

Introduction

In the modern applied science, computer simulation has become the third primary tool together with theory and experiment. A modern scientist, investigating some natural, technical, economical or even sociological phenomenon, often combines all three tools in order to understand its essence, characteristics and probable connections to other phenomena.

In order to use computer simulation properly a modern scientist needs to have some knowledge in computer science and have skills in programming. At the same time, the usage of parallel computers is becoming increasingly essential due to the requirements of large memory size, computer code run time and other factors. While writing sequential codes usually does not appear as a too complicated task for most of the scientists, parallel programming requires more special knowledge and skills. Therefore, obtaining a sufficient level of knowledge means additional time expenses and efforts, which are missing in doing deep research in the particular scientific field.

Computational Fluid Dynamics (CFD) is the fast developing field of science, where the usage of parallel computations is becoming increasingly essential due to the requirements of large memory size, computer code run time and other factors.

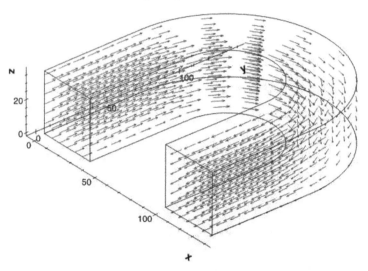

Fig. 1. An example of a calculated flow filed

Software Engineering could help in parallelization to save time and force of an applied scientist. The present work presents our positive experience using the parallelization of one particular CFD code as an example.

The problem on three-dimensional stationary ideal fluid flows through channels with complicated geometry has been investigated [1].

The realization of the numerical algorithm has led to the sequential computer code. Many test calculations using PCs have shown that the available sequential computing power is not sufficient, as we encounter the lack of computing memory, large calculation times or insufficient exactitude.

Parallel computers allow performing large amounts of computations with the required exactitude within a reasonable time scale. However, the program has to be parallelized for that. In order to save time and effort the idea appeared to use the existing parallel code as the basis for the parallelization of the considered sequential code. The mesh structure shows similarities to the mesh structure of URANUS [2], developed for the calculation of reentry flows. As the necessary communication constructs needed by the parallelization are mostly capsuled in additional modules, it should be possible to incorporate these modules into the current program. With this, we gain the communication for the solver and the residual calculation and the communication for the setup of the halo cells at program startup. In addition, the whole domain partitioning algorithm including (static) load balancing is already available.

The analysis of the sequential code has to be done in order to choose the parallelization approach, define whether shared or distributed memory should be used, specify the kind of parallelization method, make the decision about the parallelization approach for the solver, and define target computer architectures.

1 Analysis

Let us present the summary of the solution main stages and the description of the algorithm, which gives the overview of the sequential computer code under consideration. The detailed description can be found in [1].

1.1 Numerical method

- Mathematical formulation in original variables, i.e. velocity vector $\boldsymbol{u}$ and pressure p.
- Mathematical formulation in new dependent variables: vector potential $\boldsymbol{\psi}$ and vorticity vector $\boldsymbol{\omega}$.
- The transformation from Cartesian coordinates to curvilinear coordinates:

$$x^\alpha = x^\alpha(q^1, q^2, q^3), \quad \alpha = 1, 2, 3. \tag{1}$$

Though the equations became more complicated, the computational domain became simpler: i.e. the unit cube. The mesh generation is done by the equidistribution method [3]. The staggered grids are used.

- Finite difference equations for the components of the vector potential are obtained by the integro-interpolational method.
- Assuming the condition $v^2 > 0$ is fulfilled over the computational domain, the analogue of the implicit predictor-corrector scheme of S.K. Godunov [4] is applied for calculation of the components of the vorticity vector. After the iterations have converged, the components of the velocity vector can be calculated using the newly obtained components of the vector potential.

1.2 Algorithm

The following values are calculated before the iterative process starts; they do not change during iterations:

- boundary conditions for tangential components of vector potential $\boldsymbol{\psi}$;
- ω^2 on the inlet is defined by the given velocity vector $\boldsymbol{\nu}_1$.

 I. Numerical calculation of potential ideal fluid flow ($\boldsymbol{\omega} = 0$).
   1. *Local iterative process:* successive over-relaxation method (SOR) for solving difference equations for ψ_1, ψ_2, ψ_3. The iterations run for all three components simultaneously.

   The obtained solution is the initial approximation for solving the problem on vortical fluid flow. $\boldsymbol{\psi}^n$, $\boldsymbol{\omega}^n$ is the known n-th iterative approximation. The $(n+1)$-th iterative approximation is obtained in three stages.
 II. *Global iterative process:* numerical calculation of vortical ideal fluid flow.

1. *Local iterative process:* successive over-relaxation method (SOR) for solving difference equations for ψ_1, ψ_2, ψ_3. The iterations run for all three components simultaneously. ψ^{n+1} is obtained.
2. Calculation of ω^1, ω^3 on the inlet (for ω^2 see above).
3. ω^{n+1} is calculated by marching method.

   The given analysis allows understanding the structure of the sequential code, which is the basis for the parallelization.

2 Specification

In principle, there are three major opportunities to parallelize a scientific application: doing work decomposition on a loop or section basis, doing data decomposition or doing domain decomposition. Nevertheless, experience has shown that only domain decomposition can provide the desired efficiency on the largest available super computers for this class of applications.

Work decomposition needs Shared Memory platforms as a requirement and these platforms using only work decomposition show performance problems using a large number of processors [5]. Therefore, we have to aim at a domain decomposition for parallelization. As there should be no usage limit, the programming model will be message passing, so we can use platforms with shared and with distributed memory.

The programming library to be used should be MPI [6] as this is the standard for Message Passing. MPI is available on nearly all parallel platforms, especially on large and largest systems we are looking at, as the estimated run time for a real world problem with the program on a single processor system are years. The disadvantage of domain decomposition is that the whole program code has to be reviewed and it has to be adapted manually.

Initial point for the parallelization is now the calculating mesh. Decomposing the calculation mesh into the different domains, one for each process, normally results in additional communication steps, as the values of the neighbour cells, now located on a different processor, have to be made available for the local calculation (see Figure 2).

To reduce the necessary communication, we store the values from the neighbour cells locally on each process in so called halo cells, as they do not belong to the local processor. To avoid unnecessary communication, some calculation operations have to be performed for the halo cells, too, instead of communicating the result of every step. This increases the amount of calculation, but that is in general and also in our case more efficient than communicating these values, as communication is quite expensive, measured in possible floating point operations. This holds especially on clustered systems.

From the analysis step, we can identify one large problem for the parallelization, which is the SOR method for the calculation of the vector potential (see Iterative process: I.1 and II.1). There are solutions for that, e.g. the red black colouring, but these methods are not vectorizable any more. On micro-processors this means, that the cache efficiency is reduced. Tests show, that moving back to the Jacobi method does not result in the penalty expected from theory. The reason is that the convergence is not only limited by the chosen solving method, but also by the accuracy of the equations compared to the model. Therefore, the Jacobi method is chosen as solver in the parallel program.

The analysis of the whole iterative process (e.g. Chapter 2, Iterative process: II) has shown that each global iteration consists of a number of local iterations which might be quite large even more than one thousand in some cases. Within each of these local iterations, there is an equation system to solve which introduces a communication for each inner boundary to exchange the results for the halo cells. Depending on the needed accuracy, even more data exchange steps per local iteration might be necessary.

The mesh structure shows similarities to the mesh structure from another program, parallelized in a different effort [7]. As the necessary communication constructs needed by the parallelization are mostly capsuled in additional modules, it should be possible to incorporate these modules into the current program. Therefore, we form a new library out of these modules, which we will use for this parallelization. With this, we gain the communication for the solver and the residual calculation, and the communication for the setup of the halo cells at program startup. In addition, the whole domain partitioning algorithm including (static) load balancing [8] is already available.

The marching method requires that we have a dedicated dimension for this. Therefore, during domain decomposition, the dimension orientation should not be changed. The parallelization library fulfils this requirement as long as you have only one domain or domains with the same dimension orientation as input. The parallelization library itself is also able to handle so called

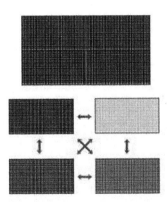

Fig. 2. Domain decomposition and communication

multiblock meshes, where the orientation on different mesh blocks might be different.

The parts of the program which need to be adapted are now mainly the data structures to incorporate the halo cells and, of cause, the handling of the boundary conditions.

In the sequential version, the boundaries are clearly defined and positioned at the six faces of the computational domain. With the domain decomposition, we now have to differentiate whether the face on the local process is a real, physical boundary or whether it is only a boundary to a neighbouring process. These boundaries, where halo cells are located, are also called inner boundaries. For the decision whether there is in an inner or a physical boundary, the available parallelization library provides some parameters to ease the programming of the boundary handling. Anyway, in the flow code, there has to be an additional part, where the inner boundaries are handled. Additionally, the parallelization library needs to have the physical boundary handling more generalized, but it has also support for that. However, the handling of the physical boundaries needs to be changed with a non-negligible manual effort.

Another critical point is I/O. Especially writing the calculated results takes a lot of time if this is not done in parallel. Therefore, we decide here to write the results calculated on each process into particular files, where the filename includes the process number. In addition, the corresponding local meshes are also written into different files. These files also include information about the particular local connectivity. Using this information, a visualization program can calculate its own view to the whole computational domain to visualize the results.

Another opportunity would be using MPI-IO [9] for reading and writing. However, this is unfortunately not available within the used parallelization library.

3 Concept

In the concept phase, we now move the considerations from the specification into a more detailed concept, which is the basis for the implementation step itself.

For the programming language, we do not have a choice as the sequential code is written in Fortran90. We cannot do anything to change that. However, most of the current supercomputers can be efficiently programmed in C and Fortran, only. In addition, from our experience we know that most Fortran programs run with a higher efficiency than equal programs written in C.

At first, we have to develop the necessary changes in the data structure for the needs of the parallel program according to the specification. This means mainly to incorporate the halo cells. The best way doing this is, to use the same position in the array for the physical boundary as for the halo cells.

With that, we can calculate the inner cells as usual without caring for any kind of boundaries or communication necessities.

In a parallel run, a specific amount of processes is started to work together, one on each requested processor, normally. The input is given in three files, one with the parameters for the calculation, one with the size of the mesh for the calculation and one with the mesh itself.

In the parallel case, it is inefficient to read this information by all processes, so we have a special process, which is caring for such particular tasks.

The first information needed for the parallelization is the mesh size, because the whole domain decomposition step is basing on this. After the domain decomposition, which is done by the parallelization library, every process knows which part of the domain it has to calculate. For this it now has to store not only the size of its local subdomain as the domain to calculate but in addition the place of this local subdomain in the global domain.

The parameters for the calculation are then read by one chosen process and broadcasted to the other processes. After that, each process reads its part of the domain including the information of its halo cells from the mesh file and starts the calculation of its part.

Within the main calculation loop, the loops over the internal part of the computational subdomains need not to be changed; only the already given parameters have different (local) values. Nevertheless, we have to adapt all loops for the boundary calculation. These loops have now to be formulated in a more general way. This means, they have to be made dimension independent.

Additionally, we have to introduce boundary conditions for the corners as it is not clear from the beginning, where which kind of corner will be.

For all the boundary loops, the parallelization library provides the required parameters. These parameters are calculated and set up during the domain decomposition step.

When parallelizing the vector potential Jacobi solving steps, after each solution step, the halo cells needs to be updated, as the values are needed in the immediately following calculation of the new local iteration. Therefore, communication is needed at that point. The parallelization library provides a subroutine.

The termination criterion for the local iterative process is a globally calculated maximum norm of the difference between the results of the current and

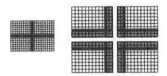

Fig. 3. Outline of halo cells (2D example)

the previous iteration. To calculate this norm in the parallel case, we have to introduce a collective communication pattern, calculating a global maximum out of the local ones of the subdomains. The MPI-call for that is also capsuled into the parallelization library, so that there is no direct call to the MPI-Library from the application program.

The handling of the marching method (Chapter 2, II.3) in the parallel program is straightforward. We do not have any dependencies of the mesh cell in the second dimension (dominating flow direction). The dependencies in the other two directions are once more nearest neighbour dependencies and we have the values of these available in the halo cells. After this, we need again a communication step to provide the changed values to the halo cells of the particular neighbours. After the calculation of the vorticity vector, we need to update the values of the neighbour cells of the particular neighbours using an additional communication step.

This communication step sets up the halo cells for the next global iteration already, as the following last step, the calculation of the Cartesian velocity components, is only necessary for a probable visualization of the flow field and does not change the values of the vector potential and the vorticity vector in the mesh cells.

Having a closer look to the communication, we have to check the communication dependencies of the particular subdomains.

An inner local subdomain has 26 neighbour subdomains. Due to the 19-point stencil we use, we have to exchange the values of the halo cells with 18 neighbours. With 6 of them we share a plane, with the other 12 we share an edge. If the subdomain is located at a plane, then we have 13 neighbours to communicate with. With 5 of them we share a plane, with the other 8 an edge. Located on an edge of the global domain, the subdomain needs values from 9 neighbours: 4 with a plane, and 5 with an edge. Being a corner, the subdomain requires a communication with 6 neighbours, 3 with a plane, and 3 with an edge.

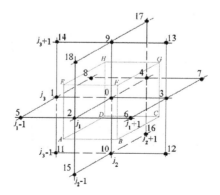

Fig. 4. The 19-point stencil

To set up the whole parallel environment and to finalize it, there are calls to MPI necessary. These are also available within the parallelization library. It is also providing a call to abort the parallel execution in an error case.

After the calculation the result files are written. For this each process writes its own output file and its local mesh file. Within the mesh file we also provide the connectivity information, which is given in a special format developed with the parallelization library. This contains all the information about the mesh block and its boundaries, so that a visualization program is able to reassemble the original domain. Anyway, a visualization program for the results of the planned large runs need to work also in parallel, as a sequential system program will not provide the memory resources necessary for calculating the pictures to visualize.

4 Implementation & test

The implementation has not been done until now, but of course, the rules for writing "good" code apply as for any sequential program. In addition, each subroutine should be tested as soon as possible, as finding errors and faults is much more complicated in the parallel case.

Unfortunately, testing like in sequential programs will not help us to remove all errors out of the parallel programs. Even going through all branches of the program will not help. The reason is that we have an influence from the outside, which is the communication with the other processes. Here, we have connections between processes that we cannot handle with the usual testing methods. For example, different timings of the processes running on different processors might be the reason that an existing error is showing up in one run, but does not in an other. Attaching a debugger to the processes might also change the timings in a way that an error does not show up any more.

The only chance to get around this would be, to run the parallel code in a way, where every possible timing delays on different process numbers are tested. However, this means to make $O(2n)$ test runs for each number of processors n. And things are even worse, as the timing can change between each communication step with in the program and so, one would need to test all these changes, too. Therefore, it is essential to deeply think about these timing constrains and the connected problems already in the concept phase and to plan the parallel constructs accordingly, in order to minimize the occurring errors.

5 Conclusion

The big advantage of the application of Software Engineering principles for that parallelization effort was the given clear structure of thinking, which for sure prevented us to make many possible mistakes and following wrong directions.

Similarly, the Software Engineering principles can be used to transform other sequential CFD codes to parallel ones with less time expenses and programming efforts.

As the necessary steps and explorations for this are likely the same in many cases it would be very helpful, if a semi automatic tool would be available to help the user with the parallelization of its code. Probably, some sort of Software Engineering tool, which supports moving from one phase to the next one with transfer functions, could help to get a good parallelization with less effort and also could help to minimize the number of occurring errors in the parallel program.

Application scientists writing their programs for sequential systems but needing the power of the largest super computers would be very pleased about such a system helping them to get a appropriate parallel code with less effort and less knowledge about parallel programming, as the average applied scientist is mainly interested in results.

References

1. Khakimzyanov GS, Shokina NYu, Küster U (2004) Numerical modeling of three-dimensional ideal fluid flows using vector potential and vorticity vector. In: Joint issue of Comp Techn 9 and Bulletin of KazNU 3/42: Proc. of the Int. Conf. "Computational and informational technologies for science, engineering and education", Almaty, Kazakhstan, part 4 (in Russian)
2. Frühauf H-H, Knab O, Daiss A, Gerlinger U (1995) J Flight Sci Space Research 19: 219-227
3. Khakimzyanov GS, Shokina NYu (1999) Russ J Numer Anal Math Model 14(4): 339-358
4. Godunov SK et al (1976) Numerical modeling of multidimensional gas-dynamics problems. Nauka, Moscow (in Russian)
5. Ciotti RB, Taft JR, Petersohn J (2000) Early experiences with the 512 processor single system image Origin 2000. In: Proc. of the 42nd CUG Conf (CUG Summit 2000) (CDROM)
6. Int J Supercomp Appl (1994) 8: 159-416
7. Bönisch TP, Rühle R (2001) Efficient flow simulation with structured multi-block meshes on current supercomputers. In: ERCOFTAC Bulletin No. 50: Parallel Computing in CFD
8. Walshaw C, Cross M, Everett M (1997) J Parallel Distrib Comput 47(2): 102-108
9. (1997) Message Passing Interface Forum: MPI-2: Extensions to the message passing interface standard. Document for a standard message-passing interface. University of Tennessee

Development of algorithm for visualization of results in scientific research

G.Balakayeva and Y.Bogdanov

Institute of Mathematics and Mechanics, al-Farabi Kazakh National University,
Masanchi str. 39/47, 480012 Almaty, Kazakhstan
balakayeva@kazsu.kz, j@dasm.kz

Nowadays computer processing of results of mathematical and numerical modelling is extremely important, and a special interest is a representation of results as diagrams, which should display the patterns of researched processes. For example, the results of numerical modelling of processes of evaporation and fuel burning in a boundary layer, the numerical solution of non isothermal filtration, numerical modelling of heat and mass transfer in reactors with a porous insertion are of a great interest for practical applications. The solution results of such problems represent a big amount of data. In order to exposure a flow pattern, it is necessary to carry out extensive computing experiments with a wide variation of parameters.

For processing scientific results there are number of graphics packages: Surfer, Grapher, Tecplot etc. A number of results of numerical modelling of the listed above problems are processed with the help of graphics processor Surfer are shown below (Fig. 1–2).

The actions expected from the researcher using the existing graphics packages, in our opinion, are inconvenient and are not common: various graphics packages need installation and adjustment of the obtained results and certain knowledge in working with them. Therefore, in our opinion the more effective point, is the development of visualization algorithms, which lead to the representation of results in real time. Then a researcher can instantly react to a program execution in numerical computing and introduce corrections into an algorithm.

The work given below offers the development of algorithm for visualization based on the sample solution of heat transfer problem. The algorithm leads to visual data presentation directly at computing process execution.

The basic mechanism of action in our algorithm will consist of allocating three memory areas for use in computing. Two of them are auxiliary at processing the initial and temporary data. The data given from the third area are used for visualization. Fig. 3 shows the block diagram of this part in the program.

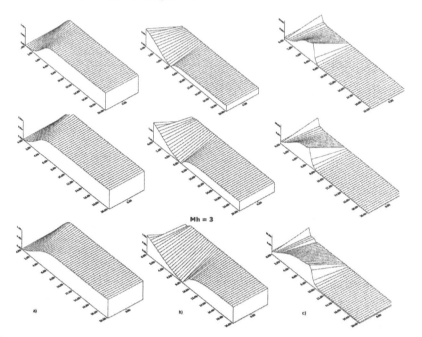

Fig. 1. Distribution of a) velocities, b) concentration, c) enthalpy in volume of a boundary layer

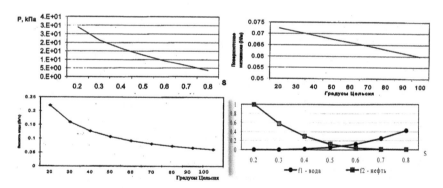

Fig. 2. Results of numerical modelling of non isothermal filtration in two phase fluid

Step 1. The beginning of the subroutine. The initialization of local variables and other necessary procedures is made here.

Step 2. Three areas of memory for the data are allocated here. The sizes of areas are determined on the previous step.

Step 3. In presence of the entrance data the condition check is made for processing. If the result of the check is positive, then there is the transition to step 4.

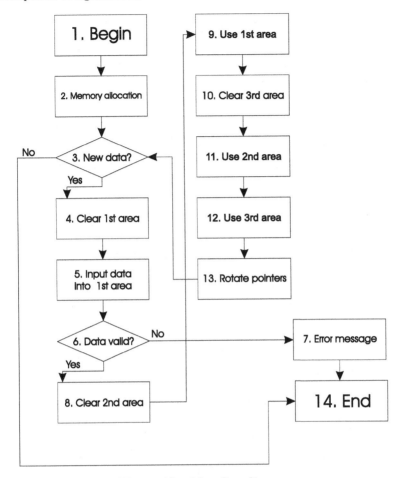

Fig. 3. Algorithm flow diagram

Step 4. The clearing of the first data area is made.

Step 5. The data from the input stream are recorded into the prepared first area.

Step 6. On the sixth step the check of integrity of the entered data and their conformity to a format accepted by the program is made. In case of any discrepancy there is the transition to the step 7, otherwise there is a transition to the step 8.

Step 7. The error message will be shown and the transition to the step 14 is done.

Step 8. The clearing of the second data area is made.

Step 9. Procedures of this step perform necessary data processing from the first area and record the result in the second data area.

Step 10. The clearing of the third data area is made.

Step 11. The procedures of this step perform necessary data processing from the second area and record the result to the third data area.

Step 12. The procedures of this step directly use the data from the third area for visualization.

Step 13. On this step there is a redefinition of indexes in all three areas of data. The pointer of the third area will point out to the first one, the pointer of the second one points out to the third one, the pointer of the first area will point out to the second area. During the rotation of the pointers the temporary variable is used so that any of working pointers would not been lost. Such a method allows to operate data much faster than a real movement of data inside memory. Besides, for a similar movement there is no necessity to reserve additional temporary memory, which is equal to a size of one area. It is enough to reserve an additional memory for temporary storing of the pointer. After performance of all procedures there is the transition to the step 3.

Step 14. The reserved memory releasing takes place here, clearing of variables and other completing procedures are done.

The results of work of the developed algorithm by the example of modelling in the mass transfer problem in a reactor are shown below in Figs. 4–5.

This program allows following the heat transfer process course from initial entrance boundary up to target boundary. The data are submitted not only in visual representation, but also the changes of numerical results are simultaneously submitted in propagation process of front temperature.

The authors now work on the adjustment of the given algorithm to complex processes following in parallel in hydrodynamics both in heat and mass transfer in non-homogenous environments.

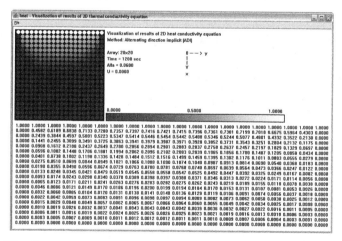

Fig. 4. Distribution of temperature in the reactor without convective members

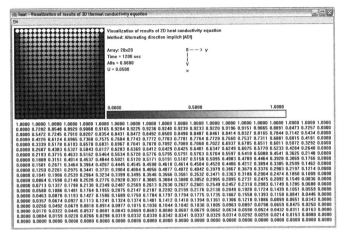

Fig. 5. Distribution of temperature in the reactor with convective members

References

1. Balakayeva GT, Yershin ShA, Zhapbasbayev UK (1999) The theory and calculation of catalytic clearing devices. Kazakh University, Almaty (in Russian)
2. Balakayeva GT, Akhimbekova AA (2001) Numerical modeling of thermal two phase filtration problem. In: Proc. of the International Conf. "Recent developments in applied mathematics and mechanics: theory, experiment and practice", devoted to the 80th anniversary of academician N.N.Yanenko. Vol.6, part 2, 41–43, Novosibirsk (in Russian)
3. Balakayeva GT (2003) Burning Plasma Chem 1(4):325–336 (in Russian)
4. Balakayeva GT (2004) Operating systems. Manual Tempus Tasis Project, Print S, Almaty (in Russian)

A general object oriented framework for discretizing non-linear evolution equations

A. Burri, A. Dedner, D. Diehl, R. Klöfkorn, M. Ohlberger

Institute of Applied Mathematics, University of Freiburg i. Br.,
Hermann-Herder-Str. 10, 79104 Freiburg i. Br., Germany
burriad,dedner,dennis,robertk,mario@mathematik.uni-freiburg.de

Summary. For a large class of non linear evolution problems we derive an abstract formulation that is based on writing the original model as a system of first order partial differential equations. Starting from this reformulation, a set of interface classes are derived that allow a problem independent implementation of various temporal and spacial discretization schemes. In particular, the abstract framework is very well suited for discretizing evolution equations with the Local Discontinuous Galerkin ansatz [1]. The implementation of the proposed framework is done within the Distributed and Unified Numerics Environment **DUNE** [2, 3].

Introduction

In [2, 3] a generic grid interface for serial and parallel computations is introduced that is realized within the Distributed and Unified Numerics Environment **DUNE**. One of the major goals of such an interface based numerics environment is the separation of data structures and algorithms. For instance, the problem implementation can be done on the basis of the interface independent of the data structure that is used for a specific application. Moreover, such a concept allows a reuse of existing codes beyond the interface. This grid interface is a central part of a number of projects with industrial and physical applications at the Institute for Applied Mathematics in Freiburg. Although a wide range of applications has to be covered, a common ground for the underlying mathematical models can be found. This observation suggests to define also a common interface for the numerical schemes and to provide a generic implementation for the central parts of the schemes. Notwithstanding the diversity of the mathematical models a discretization using the Local Dis-

---

[1] R. Klöfkorn was supported by the Bundesministerium für Bildung und Forschung under contract 03KRNCFR.

[2] M. Ohlberger was supported by the Landesstiftung Baden-Württemberg under contract 21-665.23/8.

continuous Galerkin ansatz [1] seems very promising. In this paper we describe the strategy for a general implementation of the scheme.

In the next section we outline the Discontinuous Galerkin ansatz for discretizing first order evolution equations and its extension to more complicated evolution equations. To simplify the abstract description the discretization is demonstrated in the case of a simple example. In Section 2 the projects from the Institute of Applied Mathematics in Freiburg which are using the **DUNE** code are presented, focusing on the mathematical models and their formal reformulation so that the local Discontinuous Galerkin ansatz is applicable. Section 3 is devoted to the description of the implementation details of the discretization in the **DUNE** context. We conclude with some results in Section 4.

1 The discontinuous Galerkin discretization

1.1 First order evolution equations

We begin our discussion with a study of first order systems of the form: $\partial_t \mathbf{U}(t, \cdot) = \mathcal{L}[\mathbf{U}(t, \cdot)](\cdot)$, where the spatial operator is defined by

$$\mathcal{L}[\mathbf{V}] = \mathbf{S}(\mathbf{V}) - \nabla \cdot \mathbf{F}(\mathbf{V}) - \mathbf{A}(\mathbf{V})\nabla\mathbf{V}, \tag{1}$$

where $\mathbf{V} : \mathbb{R}^d \to \mathbb{R}^p$ is in some suitable function space $\mathbb{V}$. We focus here on the space discretization, i.e., we construct a discrete operator $\mathcal{L}_h$ which maps one finite-dimensional function space V_h onto another finite-dimensional function space W_h. This operator is constructed by multiplying equation (1) with a test functions φ and by integrating over the domain $\Omega \subset \mathbb{R}^d$ which is partitioned into a finite number of cells $(\Delta_j)_{j=1,\dots,N}$. We assume that $\mathbf{V}, \varphi$ are smooth functions on the cells Δ_j but may be discontinuous over the cell interfaces. Next, we formally integrate by parts the divergence term over the cells Δ_j and we interpret the nonconservative product $\varphi\mathbf{A}(\mathbf{V})\nabla\mathbf{V}$ as a measure $(\varphi\mathbf{A}(\mathbf{V})\nabla\mathbf{V})_\Phi$ as defined in [4]. Thus we arrive at

$$\int_\Omega \mathcal{L}[\mathbf{V}]\varphi d\mathbf{x} = \sum_j \int_{\Delta_j} \mathbf{S}(\mathbf{V})\varphi d\mathbf{x} +$$

$$\sum_j \int_{\Delta_j} \mathbf{F}(\mathbf{V}) \cdot \nabla\varphi d\mathbf{x} - \sum_j \int_{\partial\Delta_j} \widetilde{\varphi\mathbf{F}(\mathbf{V})} \cdot \mathbf{n}d\sigma +$$

$$\sum_j \int_{\Delta_j} \varphi\mathbf{A}(\mathbf{V})\nabla\mathbf{V}d\mathbf{x} + \sum_j \int_{\partial\Delta_j} \widetilde{\varphi\mathbf{A}(\mathbf{V})}[\mathbf{V}]\mathbf{n}d\sigma, \tag{2}$$

where $[\mathbf{V}]\mathbf{n}$ denotes the jump of $\mathbf{V}$ in direction of the normal at cell interfaces. $\widetilde{\varphi\mathbf{A}(\mathbf{V})}[\mathbf{V}]\mathbf{n}$ is the value of the measure $(\varphi\mathbf{A}(\mathbf{V})\nabla\mathbf{V})_\Phi$ at a point of discontinuity and the averaged value $\widetilde{\varphi\mathbf{A}(\mathbf{V})}$ depends on the path Φ. The tilde denotes averaging between the cell interfaces, i.e. numerical fluxes for the conservative part. The discrete function $\mathbf{W}_h := \mathcal{L}_h[\mathbf{V}_h]$ is defined so that $\int_\Omega \mathbf{W}_h \varphi_h$ equals the right hand side of (2) for all test functions $\varphi_h \in \mathbb{W}_h$.

The construction of $\mathcal{L}_h$ depends on the averaged values used in (2), which are problem dependent. For the conservative flux average $\widetilde{\mathbf{F}(\mathbf{V})}$ one can either use a generic flux like the Lax-Friedrichs flux function or for example some more sophisticated Riemann solver based flux function [5]. In the simplest case the approximation of the measure consists of the average between the values of $\varphi_h \mathbf{A}(\mathbf{V})$ from both sides of the interface of $\partial \Delta_j$.

1.2 General systems of evolution equations

For the discretization in the case where the spatial operator $\mathcal{L}$ has a more complex form, involving for example higher order derivatives of $\mathbf{V}$ or non-linearities in the non-conservative product, we employ a decomposition of $\mathcal{L}$. Let us assume that we have spatial operators $\mathcal{L}_s$ for $s = 1, \ldots, S$ mapping $\mathbb{V}_{s-1}$ to $\mathbb{V}_s := \mathbb{W}_s \times \mathbb{V}_{s-1}$, where $\mathbb{V}_0 = \mathbb{V}$ and $\mathbb{W}_S = \mathbb{V}$ such that $\mathcal{L} = \Pi_S \circ \mathcal{L}_S \circ \cdots \circ \mathcal{L}_1$ using the projection operator $\Pi_S(\mathbb{V}_S, \ldots, \mathbb{V}_1, \mathbb{V}) := \mathbb{V}_S$. If each of the operators $\mathcal{L}_s$ is a first order operator of the form discussed above then we can construct discrete operators $\mathcal{L}_{h,i}$ as before and combine these operators to define $\mathcal{L}_h = \Pi_S \circ \mathcal{L}_{h,S} \circ \cdots \circ \mathcal{L}_{h,1}$.

Let us demonstrate this approach in the case of a simple scalar advection diffusion equation in 1d:

$$\mathcal{L}[u] = -\partial_x\big(f(u) - K(u)\partial_x(k(u))\big) - A(u)\partial_x a(u). \tag{3}$$

This equation differs in two aspects from (1); it contains a second order derivative term and the non-conservative product is non-linear in the derivative term - this leads to problems in defining the measure on the boundary [4]. We therefore formally rewrite this equation in the following form using auxiliary functions $w_1 = (w_{1,1}, w_{1,2})$:

$$w_{1,1} = \partial_x k(u),$$
$$w_{1,2} = a(u),$$
$$w_2 = -\partial_x(f(u) - K(u)w_{1,1}) - A(u)\partial_x w_{1,2}.$$

Defining the operators $\mathcal{L}_1[u] = (w_1, u)$ and $\mathcal{L}_2[w_1, u] = (w_2, w_1, u)$ we arrive at a decomposition of $\mathcal{L}$ consisting of operators of the desired form.

In some cases closure relations given for example by elliptic operators have to be taken into account. Examples are two-phase flow models and radiation hydrodynamics. In these cases we have additional terms of the form $Q(\mathcal{A}^{-1}(\mathbf{U}))$; here Q is a non-linearity and $\mathcal{A}$ some general operator. These terms can not be directly handled in the form described so far but require the construction of problem dependent discrete operators $\mathcal{A}_h^{-1}$. If these are available they can be directly included in the generic framework described above.

In this way the approach can be applied to construct discrete operators for quite complicated systems of evolution equation as we will demonstrate in Section 2.

1.3 Time discretization

So far we have a semi-implicit discretization of the evolution equation $\partial_t \mathbf{U}(t, \cdot) = \mathcal{L}[\mathbf{U}(t, \cdot)](\cdot)$ which leads to a system of ODEs $\frac{d}{dt}\mathbf{U}_h(t) = \mathcal{L}_h[\mathbf{U}_h(t)]$ for the coefficients defining $\mathbf{U}_h(t)$. To solve this system of ODEs a suitable solver, i.e., a Runge-Kutta method can be applied. Depending on the stability restrictions imposed by the spatial operator one can use either an explicit or an implicit method. To overcome time-step restrictions for explicit schemes while at the same time retaining the time accuracy of explicit methods for non-restrictive terms a suitable combination of explicit/implicit solvers is sometimes the best approach. To achieve this one has to rewrite the evolution equations using two operators

$$\frac{d}{dt}\mathbf{U}(t, \cdot) = \mathcal{L}_{\mathrm{expl}}[\mathbf{U}(t, \cdot)](\cdot) + \mathcal{L}_{\mathrm{impl}}[\mathbf{U}(t, \cdot)](\cdot), \qquad (4)$$

where $\mathcal{L}_{\mathrm{impl}}[\mathbf{U}(t, \cdot)](\cdot)$ combines all the stability restricting terms. The corresponding discrete operators are constructed in the same manner as outlined so far and for the time-discretization a semi-implicit Runge-Kutta method is used — an explicit approach for $\mathcal{L}_{\mathrm{expl},h}$ and an implicit for $\mathcal{L}_{\mathrm{impl},h}$.

Again we conclude our discussion with a simple example. Consider the evolution equation where the spatial operator is given by (3). The first order terms lead to a time-step restriction in the order of the grid size h whereas the diffusion term introduces a stability restriction of the order h^2. Therefore an appropriate splitting is given by

$$\mathcal{L}_{\mathrm{expl}}[u] = -\partial_x f(u) - A(u)\partial_x a(u), \qquad (5)$$
$$\mathcal{L}_{\mathrm{impl}}[u] = \partial_x(K(u)\partial_x k(u)). \qquad (6)$$

We use a similar decomposition as before, defining $\mathcal{L}_{\mathrm{expl},1}[u] = (w_1, u)$, $\mathcal{L}_{\mathrm{expl},2}[w_1, u] = (w_2, w_1, u)$ and $\mathcal{L}_{\mathrm{impl},1}[u] = (\hat{w}_1, u)$, $\mathcal{L}_{\mathrm{impl},2}[\hat{w}_1, u] = (\hat{w}_2, \hat{w}_1, u)$ by

$$
\begin{aligned}
w_1 &= a(u), \\
w_2 &= -\partial_x f(u) - A(u)\partial_x w_1, \\
\hat{w}_1 &= \partial_x k(u), \\
\hat{w}_2 &= \partial_x(K(u)\hat{w}_1).
\end{aligned}
$$

This leads to

$$\frac{d}{dt}\mathbf{U}(t, \cdot) = \Pi_2[\mathcal{L}_{\mathrm{expl},2}[\mathcal{L}_{\mathrm{expl},1}[\mathbf{U}(t, \cdot)]]](\cdot) + \Pi_2[\mathcal{L}_{\mathrm{impl},2}[\mathcal{L}_{\mathrm{impl},1}[\mathbf{U}(t, \cdot)]]](\cdot).$$

The simplest time-discretization is given by the forward/backward Euler method. With a time-step Δt this leads to

$$\frac{\mathbf{U}(t^{n+1},\cdot) - \mathbf{U}(t^n,\cdot)}{\Delta t} = \Pi_2[\mathcal{L}_{\text{expl},2}[\mathcal{L}_{\text{expl},1}[\mathbf{U}(t^n,\cdot)]]](\cdot) +$$
$$\Pi_2[\mathcal{L}_{\text{impl},2}[\mathcal{L}_{\text{impl},1}[\mathbf{U}(t^{n+1},\cdot)]]](\cdot).$$

We conclude that the essential steps for the discretization discussed here are the following

1. rewrite the spatial operator as a sum of two operators $\mathcal{L}_{\text{expl}}, \mathcal{L}_{\text{impl}}$.
2. decompose both spatial operators into first order operators of the form 1.
3. use the Discontinuous Galerkin approach to construct discrete operators.
4. use an explicit/implicit Runge-Kutta ODE solver to advance the solution in time.

Note that steps one and two are problem dependent whereas the following discretization steps three and four can be implemented generically if suitable numerical fluxes are available.

2 Participating projects

2.1 Torrential floods

The flow of water-sediment mixtures are often modelled starting from the incompressible two phase 3d Navier-Stokes equations for a non Newtonian fluid with free boundary. To reduce the complexity of the model a shallow flow approach is used which leads to a 2d model where the free boundary is implicitly defined by solving an equation for the depth of the flow h.

The main difficulty involves the derivation of suitable models for the internal and bed friction. Even very simple versions include very complicated terms involving both second order derivatives and non-conservative products. To demonstrate the general ideas for using the approach described previously we only present some part of the model derived in [6]:

$$\partial_t \mathbf{U} + \nabla \cdot \mathbf{F}(\mathbf{U}) - \mathbf{S}_g(\mathbf{U}) - \mathbf{S}_\tau(\mathbf{U}, \nabla \mathbf{U}, \nabla^2 \mathbf{U}) = 0. \qquad (7)$$

The balanced quantities are the depth of the flow h and the momentum $h\mathbf{u}$. $\mathbf{S}_g$ is the driving force to gravity, and $\mathbf{S}_\tau$ is a sum consisting of terms modelling the different relevant stresses. We only include two terms in the following

$$\mathbf{U} = (h, h\mathbf{u})^T,$$
$$\mathbf{F} = (h\mathbf{u}, h\mathbf{u}\mathbf{u}^T + \frac{1}{2}g_z h^2 \mathcal{I})^T,$$
$$\mathbf{S}_g = (0, g_z \nabla bh),$$
$$\mathbf{S}_\tau = (0, -\mu h \partial_x^2 u_1 + \sin(\phi_{\text{int}})g_z \text{sign}(\partial_y u_1)h\partial_y h + \ldots, \ldots).$$

Here g_z is the magnitude of gravitational acceleration, b describes the bed topology, and ϕ_{int}, μ are constants modelling the internal friction of the granular material.

With a simple transformation we can rewrite equation (7) as a system of first order equations for $\mathbf{U}$ and a set of additional quantities $\mathbf{V}_1$:

$$\mathbf{V}_1 - \nabla u_x = 0, \tag{8}$$
$$\partial_t \mathbf{U} + \nabla \cdot \mathbf{F}(\mathbf{U}) - \mathbf{S}_g(\mathbf{U}) - \mathbf{S}_2(\mathbf{U}, \mathbf{V}_1) = 0, \tag{9}$$

where the new source term is given as

$$\mathbf{S}_2(\mathbf{U}, \mathbf{V}_1) = (0, -\mu h \partial_x V_{1,1} + g_z \sin(\phi_{\text{int}}) \text{sign}(V_{1,2}) h \partial_y h + \ldots, \ldots).$$

Note the non-conservative terms in $\mathbf{S}_2$ which are linear in the derivative as required for the discretization approach described above.

2.2 Two-phase flow in porous media

In this subsection we consider two-phase multicomponent flow in porous media. Applications in mind are the simulation of water and oil flow through the soil (e.g. biodegradation) or the simulation of water and gas flow in fuel cells (see for example [7, 8, 9]).

Here, we focus on a simplified system for incompressible two-phase flow including multi-component transport which is the basis for the application in mind. The model considers gas and water flow in a porous medium where in the gas phase also the species hydrogen, oxygen, water, and some rest gas (mainly consisting of nitrogen) have to be modelled. The governing equations are given as follows. From mass and momentum balance for the water phase ($i = w$) and the gas phase ($i = g$) we get

$$\partial_t(\phi s_i) + \nabla \cdot (\mathbf{u}_i) = q_i(s_i), \tag{10}$$
$$\mathbf{u}_i = -\lambda_i(s_i) \mathbf{K} \nabla p_i,$$

with the following closure relations

$$s_w + s_g = 1, \tag{11}$$
$$p_w = p_g - p_c(s_w) = p_g - p_c(1 - s_g). \tag{12}$$

Within this model the unknown variables are the saturations of the phases s_i, the pressures of the phases p_i, and the phase velocities $\mathbf{u}_i$. ϕ denotes the porosity of the medium and $p_c(s_w)$ is the capillary pressure, given as a function of the water saturation. A parameterization for this function is for example given by the model of Van Genuchten (see [8]). $\lambda_i(s_i)$ denotes the relative mobility of the ith phase, depending on the saturation of the phases, respectively. In addition $\mathbf{K}$ denotes the absolute permeability of the medium and q_i represents sources and sinks. For instance the phase transition between the water and gas phase is modelled through q_i.

As independent variables for our computations we choose the saturation of the gas phase s_g and the pressure of the gas phase p_g. Thereby we assume, that the physics are such that we always have $s_g > 0$. Summing up the equations in (10) and using the relation (11) we get for the so called global velocity $\mathbf{u} = \mathbf{u}_g + \mathbf{u}_w$ and the global source term $q(s) = q_g(s) + q_w(1-s)$

$$\nabla \cdot \mathbf{u} = q(s_g).\tag{13}$$

The equation for p_g is determined by using relation (12) to replace p_w by $p_g - p_c(1 - s_g)$ and inserting the definitions of $\mathbf{u}, \mathbf{u}_i$ into equation (13). Thus, we get with the elliptic operator $\mathcal{A}(p_g) := -\nabla \cdot ((\lambda_w(1 - s_g) + \lambda_g(s_g))\mathbf{K}\nabla p_g)$:

$$\mathcal{A}(p_g) = q(s_g) - \nabla \cdot (\lambda_w(1 - s_g)\mathbf{K}\nabla p_c(1 - s_g)).\tag{14}$$

For each species in the gaseous phase we have to consider a transport equation. For $\mathbf{c} := \left(c^{H_2}, c^{O_2}, c^{H_2O}, c^R\right)^T$ we have

$$\partial_t(\phi s_g \mathbf{c}) + \nabla \cdot (\mathbf{u}_g \mathbf{c}) - \nabla \cdot (\phi s_g \mathbf{D}_g \nabla \mathbf{c}) = \mathbf{r}_g(s_g, s_g \mathbf{c}).\tag{15}$$

The equation for the species R can be dropped by using the relation $c^{H_2} + c^{O_2} + c^{H_2O} + c^R = 1$. Therefore the vector of the concentrations reduces to $\mathbf{c} := \left(c^{H_2}, c^{O_2}, c^{H_2O}\right)^T$. In this model $\mathbf{D}_g$ denotes the dispersion tensor describing the macroscopic diffusion of the species. $\mathbf{r}_g$ is a source/sink term which for example represents reactions of the species.

Next, we rewrite the above described model as a system of first order equations in order to apply the LDG method. Doing so, we end up with the following set of equations. The vector of unknowns is $\mathbf{U} := (\phi s_g, \phi s_g \mathbf{c})^T$ and $\mathbf{V}_1, \mathbf{V}_2$ are vectors of temporary needed values.

$$\begin{aligned}\mathbf{V}_{1,1} + \mathbf{K}\nabla p_c(1 - U_1/\phi) &= 0,\\ \mathbf{V}_{1,2} - \nabla(\mathbf{U}_2/U_1) &= 0,\end{aligned}\tag{16}$$

$$\mathbf{V}_2 + \mathbf{K}\nabla\{\mathcal{A}^{-1}(B(\mathbf{V}_{1,1}, U_1/\phi))\} = 0,\tag{17}$$

$$\partial_t \mathbf{U} + \nabla \cdot (\mathbf{F}(\mathbf{U}, \mathbf{V}_1, \mathbf{V}_2)) - \mathbf{S}(\mathbf{U}) = 0.\tag{18}$$

The first pass is formed by equations (16), the second by equations (17); here, p_g is implicitly defined by the $p_g = \mathcal{A}^{-1}(B(\mathbf{V}_{1,1}, V_{1,1}))$, where $B(\mathbf{x}, y) = q(y) - \nabla \cdot (\lambda_w(1-y)\mathbf{x})$ is the right hand side. The last pass consists of equation (18) with the flux function $\mathbf{F}$

$$\mathbf{F}(\mathbf{U}, \mathbf{V}_1, \mathbf{V}_2) := \begin{pmatrix} \lambda_g(U_1/\phi)\mathbf{V}_2 \\ \lambda_g(U_1/\phi)\mathbf{V}_2 \mathbf{U}_2/U_1 - U_1\mathbf{D}_g\mathbf{V}_{1,2} \end{pmatrix}$$

end the source/sink terms $\mathbf{S}$

$$\mathbf{S}(\mathbf{U}) := \begin{pmatrix} q_g(U_1/\phi) \\ \mathbf{r}_g(\mathbf{U}/\phi) \end{pmatrix}.$$

2.3 Reactive Navier-Stokes equations

For the simulation of reactive flows in gas turbine combustors, the reactive Navier-Stokes equations as defined in [10] are used. The mass and momentum conservation part of the compressible Navier-Stokes equations can be written as

$$\frac{\partial \rho}{\partial t} + \nabla \cdot (\rho \mathbf{u}) = 0, \tag{19}$$

$$\frac{\partial \rho \mathbf{u}}{\partial t} + \nabla \cdot (\rho \mathbf{u} \mathbf{u}^T + p\mathbf{I} - \tau) = 0. \tag{20}$$

In the momentum equation (20), τ denotes the stress tensor which for Newtonian fluids has the form

$$\tau = \mu \left(D\mathbf{u} + (D\mathbf{u})^T - \frac{2}{3}(\nabla \cdot \mathbf{u})\mathbf{I} \right), \tag{21}$$

where μ denotes the viscosity of the fluid and $\mathbf{I}$ the unity tensor.

In order to get to the reactive Navier-Stokes equations, partial differential equation for the chemical species are added:

$$\frac{\partial \rho w_i}{\partial t} + \nabla \cdot (\rho w_i \mathbf{u}) + \nabla \cdot \mathbf{j}_i = \dot{m}_i, \ i \in \{1, n_{species}\}, \tag{22}$$

where $\mathbf{j}_i$ is a flux resulting from species diffusion. The source term $\dot{m}_i$ originates from species conversions due to chemical reactions. The species diffusion is modelled using Fick's law $\mathbf{j}_i = -\rho D_i \nabla w_i - \rho \sum_{j=1}^{n_{species}} D_j \nabla w_j$, where D_i is a diffusion constant which can be chosen independently for each species i.

Due to the consideration of diffusion of species with differing enthalpy of formation ΔH_i, the standard energy equation must be supplemented with an additional term, leading to

$$\frac{\partial \rho e}{\partial t} + \nabla \cdot (\mathbf{u}(\rho e + p)) = -\nabla \cdot (\mathbf{j}_T + \tau \mathbf{u} + \sum_{i=1}^{n_{species}} \Delta H_i \mathbf{j}_i). \tag{23}$$

In this equation $\mathbf{j}_T$ is the temperature diffusion flux and is given by Fourier's law as $\mathbf{j}_T = -\lambda \nabla T$, where λ is the heat conduction coefficient and T the temperature.

In order to close the system of equations discussed here, an additional expression relating the thermodynamic variables is needed. Here, the ideal gas law $p = \rho R T$ is used, where the ideal gas constant of the mixture is expressed as $R = R^* \sum_{j=1}^{n_{species}} \frac{w_j}{W_j}$ with $R^* := 8.136 J/mol$ being the universal gas constant.

If the specific heat capacity c_v is assumed to be independent of the temperature T, the specific total energy ρe is given as

$$\rho e = \frac{1}{\gamma - 1}p + \frac{1}{2}\rho \mathbf{u}^2 + \sum_{i=1}^{n_{species}} \Delta H_i \rho w_i, \tag{24}$$

where the definition $\gamma := c_p/c_v$ and the identity $R = c_p - c_v$ is used, as well as the equation of state. Overall we arrive at a set of equations for the partial differential quantities in $\mathbf{U} = (\rho, \rho\mathbf{u}, \rho e, \rho\mathbf{w})^T$ closed using (24) to define the pressure $p = p(\mathbf{U})$ and the relation $T = \frac{p}{R\rho}$. With a simple transformation we finally arrive at a system of first order equations

$$\mathbf{V}_1 - \mu\left(D\mathbf{u} + (D\mathbf{u})^T - \frac{2}{3}(\nabla \cdot \mathbf{u})\mathbf{I}\right) = 0,$$

$$\mathbf{V}_2 + \lambda\nabla\frac{\rho R}{p(\mathbf{U})} = 0, \qquad (25)$$

$$\mathbf{V}_3 + \rho D\nabla\mathbf{w} + \rho\sum_{i=1}^{n_{spec}} D_i\nabla w_i = 0,$$

$$\partial_t\mathbf{U} + \nabla \cdot \mathbf{F}_2(\mathbf{U}, \mathbf{V}_1, \mathbf{V}_2, \mathbf{V}_3) = \mathbf{S}_2(\mathbf{U}), \qquad (26)$$

where the new flux vector $\mathbf{F}_2$ and source term $\mathbf{S}_2$ are given as

$$\mathbf{F}_2(\mathbf{U}, \mathbf{V}_1, \mathbf{V}_2, \mathbf{V}_3) = \Big(\rho\mathbf{u}, \rho\mathbf{u}\mathbf{u}^T + p(\mathbf{U})\mathbf{I} + \mathbf{V}_1,$$

$$(\rho e + p(\mathbf{U}))\mathbf{u} + \mathbf{V}_2 + \mathbf{V}_1\mathbf{u} + \sum_{i=1}^{n_{species}} \Delta H_i\mathbf{V}_{3i},$$

$$\rho\mathbf{w}\mathbf{u}^T + \mathbf{V}_3\Big)^T,$$

$$\mathbf{S}_2(\mathbf{U}) = \Big(0, 0, 0, \dot{m}(\mathbf{U})\Big)^T.$$

2.4 Liquid-vapour flows with phase-change

We consider the dynamics of a compressible liquid-vapour flow undergoing phase transitions which is governed by the compressible Navier-Stokes-Korteweg system [11]. This is an extension of the Navier-Stokes system with an additional third order term that takes the effect of surface tension at phase boundaries into account.

$$\begin{aligned}\partial_t\rho + \qquad \nabla \cdot (\rho\mathbf{u}) \qquad &= 0,\\ \partial_t(\rho\mathbf{u}) + \nabla \cdot (\rho\mathbf{u}\mathbf{u}^T) + \nabla p(\rho) &= \nabla \cdot \tau + \lambda\rho\nabla\Delta\rho,\end{aligned} \qquad (27)$$

where the unknowns are the density $\rho(\mathbf{x}, t) > 0$ and the velocity $\mathbf{u}(\mathbf{x}, t)$, τ is the usual Navier-Stokes tensor as in the section before, the capillarity $\lambda > 0$ is a known constant and the function $p = p(\rho)$ is given by a van-der-Waals equation of state where the temperature is fixed to a constant below the critical temperature to allow for phase transitions. Note that there is no order parameter in this model that indicates the phases. In this model the phases are determined by the value of the density only (low density=vapour, high density=liquid).

The system can also be written in conservative form but it turned out to be better (see [12] and references therein) to discretize the third order term together with the $\nabla p(\rho)$ term in nonconservative form in order to get accurate results when nontrivial local equilibrium configurations are present.

$$
\begin{aligned}
\partial_t \rho + \quad \nabla \cdot (\rho \mathbf{u}) \quad &= 0, \\
\partial_t (\rho \mathbf{u}) + \nabla \cdot (\rho \mathbf{u} \mathbf{u}^T) + \rho \nabla \kappa &= \nabla \cdot \tau,
\end{aligned}
\tag{28}
$$

where κ is defined by the relation

$$
\kappa = \kappa(\rho, \Delta \rho) := -\lambda \Delta \rho + \mu(\rho), \quad \mu(\rho) := \int_0^\rho \frac{p'(s)}{s} ds.
\tag{29}
$$

The advantage of the nonconservative form is that the variable κ appears explicitly in the equation. κ is equal to some constant at a static equilibrium state (i.e. a state with $\partial_t \rho = 0$ and $\mathbf{u} = \mathbf{0}$). So the discretization of the nonconservative reformulation of the equation leads naturally to a well-balanced scheme.

$$
\begin{aligned}
\mathbf{V}_\rho - \nabla \rho &= 0, \\
\mathbf{V}_\mathbf{u} - \nabla \mathbf{u} &= 0,
\end{aligned}
\tag{30}
$$

$$
\kappa - \mu(\rho) + \lambda \nabla \cdot \mathbf{V}_\rho = 0,
\tag{31}
$$

$$
\begin{aligned}
\partial_t \rho + \quad \nabla \cdot (\rho \mathbf{u}) \quad &= 0, \\
\partial_t (\rho \mathbf{u}) + \nabla \cdot (\rho \mathbf{u} \mathbf{u}^T) + \rho \nabla \kappa - \nabla \cdot \tau(\mathbf{V}_\mathbf{u}) &= 0.
\end{aligned}
\tag{32}
$$

3 Implementation of the discrete evolution operator

This section deals with the realization of the mathematical model from section 1 in **DUNE**. Subsection 3.1 describes fundamental concepts of **DUNE** relevant for the LDG ansatz whereas subsections 3.2 and 3.3 explain the newly introduced discretization concepts and their implementation.

3.1 Discrete functions and operators

The implementation of a discrete model of a partial differential equation is based on the following general concept of function spaces, functions and operators, that act on functions.

Abstract definition of function spaces and functions

A function space V in our concept is a set of mappings from the domain $D := \mathbb{K}_D^d$ to the range $R := \mathbb{K}_R^n$, e.g.

$$
V := \{ u : \mathbb{K}_D^d \to \mathbb{K}_R^n \}
$$

Here, $\mathbb{K}_D$ denotes the domain field, $\mathbb{K}_R$ the range field and d, n the dimensions of the domain and range, respectively. To further specify the function space, additional properties can be added, e.g. the function are in C^m or do belong to the Sobolev Space H^m.

A discrete function space V_h with finite dimension m is a subset of a function space with the property that the functions are defined locally on the elements e of the underlying computational grid $\mathcal{T}$. If $\hat{e}$ denotes the reference element of e and F_e the mapping $F_e : \hat{e} \rightarrow e$, we define the local base function set $V_{\hat{e}}$ on the reference element $\hat{e}$ through

$$V_{\hat{e}} := \{\varphi_1, \cdot, \varphi_{\dim(V_{\hat{e}})}\}.$$

The discrete function space V_h is then given as

$$V_h := \left\{u_h \in V : u_h|_e := u_e := \sum_{\varphi \in V_{\hat{e}}} g(u_{e,\varphi}) \, \varphi \circ F_e^{-1}, \text{ for all } e \in \mathcal{T}\right\}.$$

We call $V_e := \mathrm{span}\{\varphi \circ F_e^{-1} : \varphi \in V_{\hat{e}}\}$ a local function space, $u_e \in V_e$ a local function, and $\mathrm{DOF}_e := \{u_{e,\varphi}, \varphi \in V_{\hat{e}}\}$ the set of local degrees of freedom. In order to incorporate global properties of the discrete function space, the function space has to provide a mapping g between the local degrees of freedom (DOF_e) and the global degrees of freedom $\mathrm{DOF} := \{u_i : i = 0, \cdots, m\}$.

We summarize, that a discrete function space V_h is determined by a function space V, a grid $\mathcal{T}$, the base function sets $V_{\hat{e}}$ for all reference elements $\hat{e}$ and the mapping g from local to global degrees of freedom. A discrete function $u_h \in V_h$ is accordingly defined as a set of local functions u_e where a local function provides access to the local degrees of freedom (DOF_e).

Abstract definition of operators acting on discrete function

A discrete operator L_h is a mapping that acts on discrete functions, e.g.

$$L_h : V_h \rightarrow W_h.$$

Thereby, we suppose that a discrete operator may always be decomposed into a global Operator L_{pre}, a set of local operators L_e, and a global operator L_{post}, i.e.

$$\begin{aligned} L_{pre} &: V_h \rightarrow \{V_e, e \in \mathcal{T}\}, \\ L_e &: V_e \rightarrow W_e, \quad \text{for all } e \in \mathcal{T}, \\ L_{post} &: \{W_e, e \in \mathcal{T}\} \rightarrow W_h, \end{aligned}$$

$$L_h = L_{post} \circ \mathrm{diag}\{L_e, e \in \mathcal{T}\} \circ L_{pre}.$$

Here $\mathrm{diag}\{L_e, e \in \mathcal{T}\}$ is a diagonal matrix composed by the entries L_e. Note that with this definition of a discrete operator we are able to combine operators L_h^1 and L_h^2 in a local way, provided that $L_{pre}^2 \circ L_{post}^1 = Id$, i.e.

$$L_h^2 \circ L_h^1 = L_{post}^2 \circ \mathrm{diag}\{L_e^2 \circ L_e^1, e \in \mathcal{T}\} \circ L_{pre}^1.$$

Interface classes for discrete functions and operators

According to the abstract description of discrete functions and operators above, we define the following interface classes:

1. FunctionSpace⟨DomainField, RangeField, DomainDim, RangeDim⟩
   This class corresponds to the function space V. It is parameterized by the domain field $\mathbb{K}_D$ =DomainField, the range field $\mathbb{K}_R$ =RangeField, as well as the dimensions of the domain d =DomainDim and range n =RangeDim.

2. Function⟨FunctionSpace⟩
   A Function is parameterized by the type FunctionSpace of the function space it belongs to. To evaluate a function, the following method is provided:
   a) evaluate(x,ret): Evaluates the function at point x and returns the value ret.

3. DiscreteFunctionSpace⟨FunctionSpace, Grid, BaseFunctionSet⟩
   This class corresponds to the discrete function space V_h. It is parameterized by the type of the function space V =FunctionSpace such that $V_h \subset V$, the type of the computational grid $\mathcal{T}$ = Grid and the type of the base function set $V_{\hat{e}}$ = BaseFunctionSet. The class provides an iterator for the access of the entities e of the grid. In addition the following methods are provided:
   a) mapToGlobal(e, nLocal): Returns the global number of the degree of freedom with local number nLocal on the entity e. Thus, it corresponds to the mapping g in our abstract definition.
   b) getBaseFunctionSet(e): Returns base function set $V_{\hat{e}}$ of entity e.

4. DiscreteFunction⟨DiscreteFunctionSpace, LocalFunction⟩
   A discrete function is parameterized with the type of the discrete function space V_h = DiscreteFunctionSpace it belongs to. In addition it is also parameterized with the type of its local functions u_e LocalFunction on the entities e. To access the local functions, the following method is provided:
   a) localFunction(e, lf): Returns the local function lf of entity e.

5. DiscreteOperator⟨LocalOperator,DFDomain,DFRange⟩
   A discrete operator L_h is parameterized by the type of the functions in its domain (DFDomain) and the type of the functions in its range (DFRange). In addition the type of the local Operators L_e is given. To apply the discrete operator, the ()-operator is defined as follows:
   a) (arg, dest): Applies the operator to arg of type DFDomain and returns the resulting discrete function dest of type DFRange.

3.2 Combining discrete operators – abstract interface classes

In the following, the classes for combining discrete operators in the form required for the Local Discontinuous Galerkin (LDG) ansatz described in the previous sections are presented.

The new classes in **DUNE** are:

1. Pass⟨ProblemType, PreviousPass⟩
   As described in section 1, the operator $\mathcal{L}$ is decomposed into passes $\mathcal{L}_i$ as $\mathcal{L}[\mathbf{U}] = \mathcal{L}_K[\mathcal{L}_{K-1}[\cdots \mathcal{L}_2[\mathcal{L}_1[U]]\cdots]]$. This class serves as the base class for specialized versions of a sub operator $\mathcal{L}_i$. It controls the execution of the previous passes and assembles the data, i. e. it combines the solution of the previous passes $(\mathbf{V}_{i-1}, \ldots, \mathbf{V}_1)$ with the data $\mathbf{U}$. The actual computation of a pass is implemented in a subclass, which needs to override the method compute to this end. The class takes two template parameters:
   a) ProblemType User-defined class which provides the problem dependent types.
   b) PreviousPass Type of the previous pass implementing $\mathcal{L}_{i-1}$.
      PreviousPass is either a subtype of Pass or – in case the actual pass is the first pass – of type StartPass⟨ArgumentImp⟩.
   StartPass⟨ArgumentImp⟩ serves as an end marker to the list of passes. At run time, the structure of the passes resembles a linked list. Due to the template based implementation, the structure of this linked list can be evaluated at compile-time, so that every pass knows the exact type of its preceding passes and has access to the types defined by them.

2. PassImpl⟨ProblemType, PreviousPass⟩
   PassImpl stands for a set of generic classes which implement the evaluation of the operator $\mathcal{L}_i$ of pass i with the data obtained from the previous pass. Predefined classes exist for common cases which result from the transformation of the overall problem into a first order system. For instance, specific implementations to evaluate equations of the form (2) with a flux function $\mathbf{F}$ and a source term $\mathbf{S}$ and for solving general elliptic problems are available in **DUNE**. If the problem at hand does not fit into one of these categories, the user can define her own classes by deriving from Pass and overriding the method Pass⟨...⟩::compute.

3. ProblemImpl
   ProblemImpl stands for a set of user-defined classes which describe the actual problem. In most cases, the class consists of problem-specific type definitions and an implementation for the functions $\mathbf{S}$, $\mathbf{F}$ and $\mathbf{A}$ in equation (1). Hence, the effort to adapt the Discontinuous Galerkin Operator in **DUNE** is reduced to write a simple problem definition class for each pass while the more intricate details are shielded from the user in the predefined classes.

3.3 Combining discrete operators – example implementation

To illustrate how the classes described in section 3.2 are used, a simplified implementation is shown here. The code example implements a simplified version of the problem (1) defined in section 1, using a finite difference discretization for simplicity. The simplified form of the equation reads

$$\partial_t U(x,t) = -\partial_x \cdot (aU(x,t) + \epsilon\partial_x U(x,t)). \tag{33}$$

The core of the implementation is the class Pass. In the simplest form, an operator would contain a list of Pass pointers, sequentially evaluate them with the data and the results from the previous Passes and return the result from the last one. The solution implemented here is close to this form, with two differences:

1. There is no enclosing operator class, but the last pass serves as the operator itself. Consequently, the assembly of the correct input data to each Pass and the adherence to the right calling order is managed by the Pass class itself.

2. The list of Passes is implemented as a linked list evaluable at compile-time so that type information can be carried over from previous Passes. This is necessary since the type of the discrete functions of the data and of the results varies from Pass to Pass.

As an additional responsibility, the Pass class needs to provide storage for intermediate results which are only used within the passes. The following listing shows a simplified implementation of this core class.

```
// The Pass base class
template <class Problem, class FunctionSpace, class PreviousPass>
class Pass {
public:
  enum {passnr=PreviousPass::passnr+1};
  typedef FunctionSpace DestinationType;
  // Types from previous pass
  typedef typename PreviousPass::GlobalArgumentType GlobalArgumentType;
  typedef typename PreviousPass::NextArgumentType LocalArgumentType;
  // Types for internal usage or next pass
  typedef Pair<const GlobalArgumentType*, LocalArgumentType>
        TotalArgumentType;
  typedef Pair<DestinationType*, LocalArgumentType> NextArgumentType;
public:
  // The pass i is initialized with its precessor pass i-1. Every pass
  // triggers the allocation of temporary memory on its precessor.
  // In doing this it is guaranteed that the last pass allocates no
  // temporary memory for the result (which is passed to it by the client)
  Pass(PreviousPass& pass) :
    previousPass_(pass), dest_(0) {
    pass.initialize()
  }
  virtual ~Pass() {
    delete dest_;
    dest_ = 0;
  }
  // The application operator is called directly by the client. This way,
  // the last pass serves as the operator in the Dune sense and the memory
  // for the result (the argument dest) is passed to it directly.
  void operator() (const GlobalArgumentType& arg, DestinationType& dest)
  {
    // Evaluate the previous passes first
    previousPass_.pass(arg);
    // Build up the argument list (consisting of the data obtained
    // from the DGOperator and the results of the previous passes)
    TotalArgumentType totalArg(&arg, previousPass_.localArg());
    // Do computations related to this pass (virtual function)
    compute(totalArg, dest);
  }
```

```
// Trigger the allocation of temporary memory
virtual void initialize () {}
private :
  // The functions here are not part of the public interface but must be
  // accessible by all other passes
  template <class P, class PP, class PPP>
  friend class Pass;
  // Pass is called from preceding pass. The only difference to the
  // application operator is that the use of this function causes the pass
  // to use its own temporary discrete function instead of a given one
  void pass(const GlobalArgumentType& arg) {
    operator()(arg, *dest_);
  }
  // Return a list of the results of all previous passes including this one
  NextArgumentType localArg() {
    return NextArgumentType(dest_, previousPass_.localArg());
  }
private :
  // The actual computations are delegated to a subclass
  virtual void compute(const TotalArgumentType& arg,
                       DestinationType& dest) = 0;
  PreviousPass& previousPass_;
protected :
  // Temporary storage. Subclass must initialize it when needed.
  mutable DestinationType* dest_;
};
```

The implementation of the method `compute` is deferred to a subclass in order
to allow for different implementations, tailored to the user's need. Moreover,
`Pass` hides the intricate details and compile-time constructs from the user
and from the implementor of specific passes. The following listing shows an
example of such a derived `Pass` class. The `FDPass` class implements the finite
difference evaluation of a numerical flux g, defined by its `Problem` template
argument. The `compute` method consists of an iteration over the grid which
calls the numerical flux on each grid node and returns the resulting update
vector to the caller. Note that `FDPass` itself is a generic implementation that
works for a wide range of problem definitions, which the user can express by
writing a respective `Problem` class.

```
// Generic implementation of a first order FD scheme
// Problem must provide numeric flux function g
template <class Problem, class FunctionSpace, class PreviousPass>
class FDPass : public Pass<Problem, FunctionSpace, PreviousPass> {
public :
  typedef Pass<Problem, FunctionSpace, PreviousPass> BaseType;
  typedef typename BaseType :: TotalArgumentType ArgumentType;
  typedef FunctionSpace DestinationType;
  enum { passnr=Pass<Problem, FunctionSpace, PreviousPass >:: passnr };
public :
  // Constructor initializes base class and creates storage
  FDPass(PreviousPass& prevPass, Problem& prob, int pN) :
    problem(prob), BaseType(prevPass), gridsize_(pN) {}
  // Builds up temporary storage for pass (the V_i)
  virtual void initialize () {
    this->dest_=new DestinationType(gridsize_*Problem :: range);
  }
private :
  typename Problem :: ConsType rflux ,w;
  // Compute provides the actual evaluation of the operator belonging to
  // this pass. arg is a list containing the argument data
  // as well as the results
  virtual void compute(const ArgumentType& arg, DestinationType& dest) {
```

```
dest.clear();
double h=1./double(dest.size());
StateVector<Problem::domain> xl,xr;
xl[0]=0.;
xr[0]=1.;
w.clear();
for (int i=0;i<dest.size();++i) {
  if (i<dest.size()-1)
    problem.g(i,i+1,xr,xl,arg,rflux);
  else
    rflux.clear();
  w-=rflux;
  w/=(-h);
  dest.set(i,w);
  w=rflux;
  }
 }
 Problem &problem;
 int gridsize_;
};
```

The next listing shows how the problem can be put into code. The class
Problem implements the scalar linear transport problem with an additional
diffusion term from equation (33). Note that this is the only part the user
must write. Very little information about the other passes is necessary, namely
which components of the argument type contain the relevant information. To
simplify the usage, the problem class acts as an operator itself by just for-
warding the call to the last pass.

```
// The problem class organizes the passes and provides the functionality
// of an operator which can be used in a time-integrator.
class Problem {
 private:
  // Problem class for diffusion-pass dx (eps*v)
  class ProblemDiffusion {
  public:
    enum {domain=1};
    enum {range=1};
    typedef StateVector<range> ConsType;
    typedef DiscreteFunction<domain, range> DestinationType;
    ProblemDiffusion(double eps) :
      koeff(eps) {}
    // This method implements the numerical diffusion flux d_xx u
    template <class ArgumentType>
    void g(int eleft,int eright,
           StateVector<domain> xl,StateVector<domain> xr,
           const ArgumentType& arg, ConsType& dest) {
      ConsType ul;
      Element<0>::get(arg)->evaluate(eleft,xl,ul);
      dest=koeff*ul;
    }
  private:
    double koeff;
  };
  // Problem class for Transport-pass dx (a*u-v)
  class ProblemTransport { // dx (a*u-v)
  public:
    enum {domain=1};
    enum {range=1};
    typedef StateVector<range> ConsType;
    typedef DiscreteFunction<domain, range> DestinationType;
    ProblemTransport(double a) : koeff(a) {}
```

```
  // Implementation of the numerical (upwind) flux
  template <class ArgumentType>
  void g(int eleft, int eright,
        StateVector<domain> xl, StateVector<domain> xr,
        const ArgumentType& arg, ConsType& dest) {
    ConsType ul, ur;
    Element <0>:: get(arg)->evaluate(eleft, xl, ul);
    Element <0>:: get(arg)->evaluate(eright, xr, ur);
    if (koeff >0)
      dest=koeff*ul;
    else
      dest=koeff*ur;

    ConsType vr;
    At<diffpass >:: get(arg)->evaluate(eright, xr, vr);
    dest -=vr;
  }
private:
  double koeff;
};
public:
// Argument/Destination type for operator
typedef DiscreteFunction <1, 1> GlobalArgumentType;
typedef GlobalArgumentType GlobalDestinationType;
// Constructor.
// The constructor of the problem initializes all problems and passes
Problem(double a, double eps, int N) :
  N_(N),
  problem1(eps), problem2(a),
  pass1(pass0, problem1,N),
  pass2(pass1, problem2,N) {
}
// The problem provides its own application operator and acts as a Dune
// operator in its own right.
void operator() (const GlobalArgumentType& arg,
                GlobalDestinationType& dest) {
  pass2(arg, dest);
}

// Size of the grid
int size() {
  return N_;
}
private:
  typedef PassStart<GlobalArgumentType> PassStartType;
  typedef FDPass<ProblemDiffusion, DiscreteFunction <1, 1>, PassStartType>
    Pass1Type;
  typedef FDPass<ProblemTransport, DiscreteFunction <1, 1>, Pass1Type>
    Pass2Type;
  int N_;
  ProblemDiffusion problem1;
  ProblemTransport problem2;
  PassStartType pass0;
  Pass1Type pass1;
  Pass2Type pass2;
  enum {diffpass=Pass1Type:: passnr, transpass=Pass2Type:: passnr};
};
```

Putting it all together consists of instantiating the problem class and initializing the correct data. The evaluation of the operator finally is as simple as writing prob(data, sol), as can be seen in the next listing. Here, the information about the time-stepping scheme is handled in an additional class (ForwardEuler) not presented in this paper.

```
int main(int argc, char ** argv) {
  int N=atoi(argv[1]); // number of grid nodes
  double T=atof(argv[2]); // end time
  double eps=atof(argv[3]); // viscosity
  double a=atof(argv[4]); // advection velocity
  // Creation of the problem and the corresponding operator
  Problem prob(a,eps,N);
  // Creation of the data vectors
  Problem::GlobalArgumentType u0(N), u(N);
  // Creation of the timestepping operator (not defined here)
  ForwardEuler<Problem> rk(prob,0.);
  // Initialization of the data
  initData(u, u0);
  //Time loop
  while (rk.time()<T) {
    rk(u,u);
    std::cerr << rk.time() << std::endl;
  }
  // Output result
  outputResult(u);
}
```

4 Results and conclusion

We apply the method to the Navier-Stokes-Korteweg system described in Subsection 2.4. For definition of the averaged values and numerical fluxes we refer to [12]. Time discretization is done by application of implicit Runge-Kutta methods.

As test case for the method we choose an initial configuration in two space dimensions for which the solution of the Navier-Stokes-Korteweg system is *quasi*-known. This means the solution is known to exist and can be computed very accurately by a different scheme. Figure 1 shows the result of this computation. The left part of the figure shows that the expected order of the method will be reached, the right figure demonstrates the efficiency of the method i.e. higher order methods lead to more efficient methods (provided that the solution is smooth).

References

1. Cockburn B, Shu C-W (1998) SIAM J Numer Anal 35: 2440–2463
2. Bastian P, Droske M, Engwer C, Klöfkorn R, Neubauer T, Ohlberger M, Rumpf M (2004) Towards a unified framework for scientific computing. In: Kornhuber R, Hoppe R, Périaux J, Pironneau O, Widlund O, Xu J (eds) Domain decomposition methods in science and engineering. Springer, Berlin Heidelberg New York
3. Burri A, Dedner A, Klöfkorn R, Ohlberger M (2005) An efficient implementation of an adaptive and parallel grid in DUNE. Tech. rep., submitted to: Proc. of the 2nd Russian-German Advanced Research Workshop on Computational Science and High Performance Computing
4. Dal Maso G, LeFloch P, Murat F (1995) J Math Pures Appl 74: 483–548

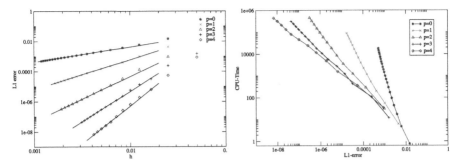

Fig. 1. DG approximation for ansatz functions with different polynomial degree p. Mesh size h versus L^1-error (the black lines indicate the expected order) and L^1-error versus CPU-time

5. LeVeque R (1990) Numerical methods for conservation laws. Lectures in Mathematics, Birkhäuser, first edn
6. Vollmöller P (2004) J Comput Phys 199(1):150–174
7. Bastian P, Rivière B (2004) Discontinuous galerkin methods for two-phase flow in porous media. Tech. Rep. 2004–28, IWR (SFB 359), Universität Heidelberg
8. Helmig R (1997) Multiphase flow and transport processes in the subsurface: a contribution to the modeling of hydrosystems. Springer.
9. Kühn K, Ohlberger M, Schumacher J, Ziegler R, Klöfkorn R (2003) A dynamic two-phase flow model of proton exchange membrane fuel cells. Preprint CSCAMM Report 03-07, submitted to the 2nd EUROPEAN PEFC FORUM, Luzern
10. Poinsot T, Veynante D (2001) Theoretical and numerical combustion. R.T. Edwards, Inc.
11. Anderson D, McFadden G, Wheeler A (1998) Ann Rev Fluid Mech 30: 139–165
12. Diehl D (2005) Well balanced discontinuous Galerkin schemes for the Navier-Stokes-Korteweg equations. In: Proc. of the Conf.: Hyperbolic Problems: Theory, Numerics, Applications Osaka 2004, Yokohama Publishers, Inc

The Cauchy problem for Laplace equation on the plane

S.I. Kabanikhin[1,2] and G. Dairbaeva[3]

[1] Sobolev Institute of Mathematics SB RAS Lavrentiev Ave. 4 Novosibirsk, 630090, Russia
[2] Kazakh–British Technical University Tolebi ave. 59, 050091 Almaty, Kazakhstan kabanikh@math.nsc.ru
[3] al-Farabi Kazakh National University Al-Farabi av. 71, 050078 Almaty, Kazakhstan dairbaeva@kazsu.kz

1 Introduction

Let Ω be a bounded simply connected domain in $R^2 = \{(x, y)\}$ with a continuous piece-smooth boundary $\Gamma = \partial\Omega$. Let Γ be divided into two connected parts $\Gamma_1 \cup \Gamma_2 = \Gamma$, $\Gamma_1 \cap \Gamma_2 = \emptyset$. Besides we shall assume that straight lines parallel to the coordinates axes cross the boundary $\partial\Omega$ no more than at two points.

Consider the following problem

$$\Delta u = 0, \quad (x, y) \in \Omega, \tag{1}$$

$$u\bigg|_{\Gamma_1} = f(x, y), \quad (x, y) \in \Gamma_1, \tag{2}$$

$$\frac{\partial u}{\partial n}\bigg|_{\Gamma_1} = 0. \tag{3}$$

The problem (1)-(3) is ill-posed according to Hadamard. The solution is unique, but it is not stable with respect to a small perturbations of the function f. That is why the direct application of numerical methods to the problem (1)-(3) leads to the problem of error accumulation. To avoid the difficulties mentioned, we use the solutions of steady boundary problems for the same differential equation (1) instead of solving the ill-posed problem (1)-(3). Such idea was proposed by Kabanikhin S.I. and Karchevskiy A. L. [1], and developed by Kabanikhin S.I., Bektemesov M.A., Ayapbergenova A.T. and Nechayev D.V. [2].

As a well-posed problem for (1)-(3), for instance, the Neumann problem for the initial Laplace equation in the domain Ω may be used, which requires knowledge of boundary values of a normal derivative of the solution on the whole boundary. Though according to the condition (3), we are only given boundary conditions on a part of the boundary. That is why it is necessary to associate boundary values of the initial function f with the boundary values of the normal derivative of the solution on the additional part of the boundary that is quite a distinct problem.

Researching properties of the operator connecting f with the above mentioned boundary values of the normal derivative of the solution and its conjugate operator, allows to use variational method to look for an approximate solution of the initial problem (1)-(3).

Analogous approach for other problems may be found in [2], [3].

In this paper justification of the use of the steepest descent method is given for solving the problem (1)-(3).

2 The reduction of the initial problem (1)-(3) to some inverse problem

In the domain Ω, consider the Neumann problem for the Laplace equation

$$\Delta \widetilde{u} = 0, \quad (x, y) \in \Omega, \tag{4}$$

$$\left.\frac{\partial \widetilde{u}}{\partial n}\right|_{\partial \Omega} = \begin{cases} 0, & (x, y) \in \Gamma_1, \\ q(x, y), & (x, y) \in \Gamma_2. \end{cases} \tag{5}$$

Introduce the new function $u = \widetilde{u} - \dfrac{1}{mes\Omega} \displaystyle\int_{\Omega} \widetilde{u}\,dxdy$, then we will have the same problem

$$\Delta u = 0, \quad (x, y) \in \Omega, \tag{6}$$

$$\left.\frac{\partial u}{\partial n}\right|_{\partial \Omega} = \begin{cases} 0, & (x, y) \in \Gamma_1, \\ q(x, y), & (x, y) \in \Gamma_2, \end{cases} \tag{7}$$

whose solution $u(x, y)$ possesses the property

$$\int_{\Omega} u(x, y)\,dxdy = 0. \tag{8}$$

Further, the problem (6), (7) with the condition (8) will be called direct.

Suppose the function $q(x, y)$ is not known, but for solving the problem (6), (7) with the condition (8), additional information is given

$$u\Big|_{\Gamma_1} = f(x, y), \quad (x, y) \in \Gamma_1. \tag{9}$$

Using the function $f(x, y)$ it is required to find the function $q(x, y)$. The conditions on f and q are given below.

Thus, the initial problem (1)-(3) has been reduced to the inverse problem in regard to the well-posed direct problem (6), (7) with the condition (8).

Using the equations (6), (7), (9) define the operator

$$A: \ q := \left.\frac{\partial u}{\partial n}\right|_{\Gamma_2} \mapsto f := u|_{\Gamma_1}, \tag{10}$$

where $u(x, y)$ is the solution of the problem (6), (7) with the condition (8). Then the inverse problem will be written in the operator form

$$A(q) = f. \tag{11}$$

3 Researching the generalized solution of the direct problem

The function $u \in L_2(\Omega)$ satisfying the integral identity

$$\int_{\Omega} u \Delta \omega dx dy + \int_{\Gamma_2} \omega q ds = 0 \tag{12}$$

will be called the generalized solution of the direct problem (6), (7) for all $\omega \in H^2(\Omega)$ such that

$$\left.\frac{\partial \omega}{\partial n}\right|_{\partial \Omega} = 0. \tag{13}$$

Choosing a special class of the generalized solutions allows to determine the operator A in a possibly wide function class.

If $u \in C^2(\Omega)$, then the generalized solution of the problem (6), (7) is a classical one.

Let's show that the direct problem (7), (8) is well-posed in contrast to the initial one (1)-(3).

Theorem 1. *(Existence of the generalized solution of the direct problem).*

Let $q \in L_2(\Gamma_2)$ and $\displaystyle\int_{\Gamma_2} q(x, y) ds = 0$. Then the problem (6), (7) has the unique generalized solution $u \in L_2(\Omega)$ satisfying the condition (8) and for it the estimation

$$\|u\|_{L_2(\Omega)} \le C\|q\|_{L_2(\Gamma_2)}, \tag{14}$$

is fulfilled, where C is a positive constant which doesn't depend on q.

Proof. Consider $\bar{q} \in C_0^\infty(\Gamma_2)$, then the solution of the problem (6), (7) $\bar{u} \in C^\infty(\Omega)$ exists and uniquely for the condition (8) and continuously depends on the initial data [4].

Consider the auxiliary problem

$$\Delta\omega = \bar{u}, \quad (x,y) \in \Omega, \tag{15}$$

$$\left.\frac{\partial\omega}{\partial n}\right|_{\partial\Omega} = 0. \tag{16}$$

In the following integral the Green formula will be used and the boundary condition (16) will be taken into account

$$\int_{\Omega} \omega\bar{u}dxdy = \int_{\Omega} \omega\Delta\omega dxdy = -\|\nabla\omega\|_{L_2(\Omega)}^2.$$

Hence the inequality

$$\|\nabla\omega\|_{L_2(\Omega)}^2 \le \|\omega\|_{L_2(\Omega)} \cdot \|\bar{u}\|_{L_2(\Omega)} \tag{17}$$

follows.

For the solution ω of the problem (15), (16) there is the estimation [4]

$$\|\omega\|_{L_2(\Omega)} \le C_1\|\bar{u}\|_{L_2(\Omega)} \tag{18}$$

takes place, where C_1 is a positive constant which doesn't depend on $\bar{u}$. Then using the ratios (17), (18) we get

$$\|\omega\|_{H^1(\Omega)} \le C_1\|\bar{u}\|_{L_2(\Omega)}. \tag{19}$$

In the equality (12), the solution of the problem (15), (16) will be used as ω, and instead of u, q $\bar{u}, \bar{q}$ will be taken respectively, then it will be easy to get the estimation

$$\|\bar{u}\|_{L_2(\Omega)}^2 \le \|\omega\|_{L_2(\Gamma_2)} \cdot \|\bar{q}\|_{L_2(\Gamma_2)} \quad . \tag{20}$$

The inclusion inequality takes place [5]

$$\|\omega\|_{L_2(\partial\Omega)} \le C_2\|\omega\|_{H^1(\Omega)}, \tag{21}$$

where C_2 is a positive constant, which doesn't depend on ω. Next, from the inequalities (20), (21) and (19) we shall obtain

$$\|\bar{u}\|_{L_2(\Omega)} \le C_3\|\bar{q}\|_{L_2(\Gamma_2)}, \tag{22}$$

where $C_3 = C_1C_2$.

Let the sequence $q_n \in C_0^\infty(\Gamma_2)$ is such that $\displaystyle\int_{\Gamma_2} q_n ds = 0$, and it converges to $q \in L_2(\Gamma_2)$, where $\displaystyle\int_{\Gamma_2} qds = 0$, then the sequence of the solutions $u_n \in C^\infty(\Omega)$ corresponding to q_n, converges to the solution $u \in L_2(\Omega)$. Then in virtue of continuity of a scalar production, the estimation (14) with the constant $C = C_3$ will be being fulfilled.

Now one may proof limitedness of the operator A.

Theorem 2. *Let the conditions of the Theorem 1 hold true, then the solution of the direct problem has the trace* $u|_{\Gamma_1} \in L_2(\Gamma_1)$ *and the estimation*

$$\|u\|_{L_2(\Gamma_1)} \leq C\|q\|_{L_2(\Gamma_2)}, \tag{23}$$

is being fulfilled, where C *is a positive constant which doesn't depend on q.*

Proof. Let $\bar{q} \in C_0^\infty(\Gamma_2)$ and $\int_{\Gamma_2} \bar{q}ds = 0$, then the solution of the problem (6), (7) $\bar{u} \in C^\infty(\Omega)$ exists and uniquely for the condition $\int_{\Omega} \bar{u}dxdy = 0$ and continuously depends on data. Multiply the equation (6) by $\bar{u}$ and integrate in Ω, using Green's formula

$$0 = \int_{\Omega} \bar{u}\Delta\bar{u}dxdy = \int_{\partial\Omega} \bar{u}\frac{\partial\bar{u}}{\partial n}ds - \int_{\Omega}(\nabla\bar{u})^2dxdy =$$

$$= \int_{\Gamma_2} \bar{u}\bar{q}ds - \|\nabla\bar{u}\|_{L_2(\Omega)}^2.$$

Then

$$\|\nabla\bar{u}\|_{L_2(\Omega)}^2 \leq \|\bar{u}\|_{L_2(\Gamma_2)} \cdot \|\bar{q}\|_{L_2(\Gamma_2)} \leq \|\bar{u}\|_{L_2(\partial\Omega)} \cdot \|\bar{q}\|_{L_2(\Gamma_2)}. \tag{24}$$

Next, majorize the right part of the estimate (24) with the help of (21) and the Cauchy inequality with $\varepsilon > 0$

$$\|\nabla\bar{u}\|_{L_2(\Omega)}^2 \leq C_2\|\bar{u}\|_{H^1(\Omega)} \cdot \|\bar{q}\|_{L_2(\Gamma_2)} \leq$$

$$\leq \varepsilon\left(\|\bar{u}\|_{L_2(\Omega)}^2 + \|\nabla\bar{u}\|_{L_2(\Omega)}^2\right) + C_\varepsilon\|\bar{q}\|_{L_2(\Omega)}^2,$$

where $C_\varepsilon = C_2^2/4\varepsilon$. Hence, selecting $\varepsilon = \dfrac{1}{2}$ and using the estimate (14), we have

$$\|\nabla\bar{u}\|_{L_2(\Omega)} \leq C_4\|\bar{q}\|_{L_2(\Gamma_2)},$$

where $C_4 = \sqrt{C_2^2 + C_3^2}/2$. Further, doing analogically, as in Theorem 1, we shall obtain that for the solution of the direct problem (6), (7) the inequality takes place

$$\|\nabla u\|_{L_2(\Omega)} \leq C_4\|q\|_{L_2(\Gamma_2)}. \tag{25}$$

Then from the inclusion inequality (21) in virtue of (25) and Theorem 1 we shall get the required estimate

$$\|u\|_{L_2(\Gamma_1)} \leq C_5\|q\|_{L_2(\Gamma_2)},$$

where $C_5 = C_2(C_3 + C_4)$, and C_2, C_3 are identified by the ratios (21), (22).

Remark 1. *From Theorem 2 it follows that the operator A has the form*

$$A : \left\{ q \Big| q \in L_2(\Gamma_2), \int_{\Gamma_2} q ds = 0 \right\} \mapsto \left\{ f \Big| f \in L_2(\Gamma_1) \right\}$$

and for its norm the following estimate $\|A\| \leq C_5$ *takes place.*

4 Investigation of the generalized solution of the conjugate problem

In the given point a conjugate operator is calculated to the operator A.
    Consider the following Neumann problem for the Laplace equation.

$$\Delta \psi = 0, \quad (x, y) \in \Omega, \tag{26}$$

$$\frac{\partial \psi}{\partial n}\Big|_{\partial \Omega} = \begin{cases} \mu(x, y), & (x, y) \in \Gamma_1, \\ 0, & (x, y) \in \Gamma_2. \end{cases} \tag{27}$$

As the generalized solution of the problem (26), (27) will be called the function $\psi \in L_2(\Omega)$ satisfying the integral identity

$$\int_{\Omega} \psi \Delta v dx dy + \int_{\Gamma_1} v(x, y) \mu(x, y) ds = 0 \tag{28}$$

for all $v \in H^2(\Omega)$, $\frac{\partial v}{\partial n}\Big|_{\partial \Omega} = 0$.

From Remark 1 it follows that the operator A defined by the formula (10) has no dense domain of determination in $L_2(\Gamma_2)$. In fact, it is defined only for the functions $q \in L_2(\Gamma_2)$ orthogonal to one, that is, $\int_{\Gamma_2} q ds = 0$. Though this circumstance will non interfere with our defining a conjugate operator to the operator A in a subspace of the space $L_2(\Gamma_2)$.

Let ψ is a solution of the problem (26), (27), and u is a solution of a direct problem (6), (7). Multiply the equation (6) by ψ and integrate in Ω, using the Green formula, as a result the equality

$$\int_{\Gamma_1} f \frac{\partial \psi}{\partial n} ds = \int_{\Gamma_2} \psi q ds \tag{29}$$

will be obtained.

As $\int_{\Gamma_2} q ds = 0$, the right part (29) is fulfilled for $\psi|_{\Gamma_2}$ accurate to the constant

$$\int\limits_{\Gamma_1} f\frac{\partial\psi}{\partial n}ds = \int\limits_{\Gamma_2} (\psi + C)qds,$$

where C is const. Subordinate the choice of a constant to the condition $\int\limits_{\Gamma_2} \tilde{\psi}ds = 0$, where $\tilde{\psi} = \psi + C$, that is, $C = -\dfrac{1}{mes\Gamma_2}\int\limits_{\Gamma_2} \psi ds.$

Then according to the formula

$$A^* : \quad \mu := \frac{\partial\psi}{\partial n}\bigg|_{\Gamma_1} \mapsto h := \psi|_{\Gamma_2}, \tag{30}$$

where $\int\limits_{\Gamma_2} \psi ds = 0$, a conjugate operator to A is defined uniquely.

If $\psi \in C^2(\Omega)$, the generalized solution of the problem (26), (27) will be classical.

Theorem 3. *(Existence of the generalized solution of the problem (26), (27)).* *Let $\mu \in L_2(\Gamma_1)$ and $\int\limits_{\Gamma_1} \mu ds = 0$, then the problem (26), (27) has the unique generalized solution $\psi \in L_2(\Omega)$ satisfying the equality $\int\limits_{\Omega} \psi dxdy = 0$ and the following estimate*

$$\|\psi\|_{L_2(\Omega)} \le C\|\mu\|_{L_2(\Gamma_1)} \tag{31}$$

is fulfilled, where C is a positive constant, which does not depend on μ.

Proof. Let $\bar{\mu} \in C_0^\infty(\Gamma_1)$ and $\int\limits_{\Gamma_1} \bar{\mu}ds = 0$, then the solution of the problem (26), (27) $\bar{\psi} \in C^\infty(\Omega)$ exists and uniquely for the condition $\int\limits_{\Omega} \bar{\psi}dxdy = 0$ and continuously depends on data [4].

Consider an auxiliary problem

$$\Delta v = \bar{\psi}, \quad (x,y) \in \Omega, \tag{32}$$

$$\frac{\partial v}{\partial n}\bigg|_{\partial\Omega} = 0. \tag{33}$$

Multiply the equation (32) by v and integrate in Ω, taking into account the boundary condition (33)

$$\int\limits_{\Omega} v\bar{\psi}dxdy = \int\limits_{\Omega} v\Delta v dxdy = -\|\nabla v\|^2_{L_2(\Omega)}.$$

Hence

$$\left\|\nabla v\right\|_{L_2(\Omega)}^2 \leq \left\|v\right\|_{L_2(\Omega)} \cdot \left\|\bar{\psi}\right\|_{L_2(\Omega)}. \tag{34}$$

For the solution of the problem (32), (33) the estimate

$$\left\|v\right\|_{L_2(\Omega)} \leq C_6\left\|\bar{\psi}\right\|_{L_2(\Omega)} \tag{35}$$

takes place, where C_6 is a positive constant which doesn't depend on $\bar{\psi}$.
    Then from (34), (35)

$$\left\|v\right\|_{H^1(\Omega)} \leq C_7\left\|\bar{\psi}\right\|_{L_2(\Omega)} \tag{36}$$

follows, where $C_7 = \sqrt{C_6(1+C_6)}$.
    In the identity (28) take $\bar{\psi}$, $\bar{\mu}$ instead of ψ, μ respectively, and as the function v take the solution of the problem (32)-(33), as a result we have

$$\left\|\bar{\psi}\right\|_{L_2(\Omega)}^2 \leq \left\|v\right\|_{L_2(\Gamma_1)} \cdot \left\|\bar{\mu}\right\|_{L_2(\Gamma_1)} \leq \left\|v\right\|_{L_2(\partial\Omega)} \cdot \left\|\bar{\mu}\right\|_{L_2(\Gamma_1)}.$$

Then using the inequalities (21), (36) we shall get

$$\left\|\bar{\psi}\right\|_{L_2(\Omega)} \leq C_8\left\|\bar{\mu}\right\|_{L_2(\Gamma_1)}, \tag{37}$$

where $C_8 = C_2 C_7$.
    Let $\mu \in L_2(\Gamma_1)$, $\int_{\Gamma_1} \mu ds = 0$ and the sequence $\mu_n \in C_0^\infty(\Gamma_1)$, $\int_{\Gamma_1} \mu_n ds = 0$ converges to μ for $n \to \infty$ in the norm $L_2(\Gamma_1)$. Then the sequence of the solutions $\psi_n \in C^\infty(\Omega)$, $\int_\Omega \psi_n dxdy = 0$ of the problem (26), (27) corresponding to the sequence μ_n will converge to the solution of $\psi \in L_2(\Omega)$, $\int_\Omega \psi dxdy = 0$ in virtue to the continuity of the scalar production, and from the estimate (37) we get the required estimate (31) with the constant $C = C_8$.

Theorem 4. *Let the conditions of the Theorem 3 hold true. Then the solution of the problem (26), (27) has the trace $\psi\big|_{\Gamma_2} \in L_2(\Gamma_2)$ and the estimate*

$$\left\|\psi\right\|_{L_2(\Gamma_2)} \leq C\left\|\mu\right\|_{L_2(\Gamma_1)}, \tag{38}$$

is fulfilled, where C is a positive constant which doesn't depend on μ.

    Proof. Let $\bar{\mu} \in C_0^\infty(\Gamma_1)$, $\int_{\Gamma_1} \bar{\mu} ds = 0$, and $\bar{\psi} \in C^\infty(\Omega)$ is a corresponding solution of the problem (26), (27) and $\int_\Omega \bar{\psi} dxdy = 0$.
    Consider the equation (26) under $\psi = \bar{\psi}$, multiply it by $\bar{\psi}$ and integrate in Ω, as a result we shall get

$$\|\nabla\psi\|_{L_2(\Omega)}^2 = \int_{\Gamma_1} \bar{\psi}\,\bar{\mu}ds.$$

Hence

$$\|\nabla\bar{\psi}\|_{L_2(\Omega)}^2 \leq \|\bar{\psi}\|_{L_2(\Gamma_1)} \cdot \|\bar{\mu}\|_{L_2(\Gamma_1)} \leq \|\bar{\psi}\|_{L_2(\partial\Omega)} \cdot \|\bar{\mu}\|_{L_2(\Gamma_1)}.$$

Further, acting analogously, as it has been done while calculating the estimate for the norm $\|\nabla\bar{u}\|_{L_2(\Omega)}^2$ in Theorem 2, and using the estimate (31) under $C = C_{8,}$, we have

$$\|\nabla\bar{\psi}\|_{L_2(\Omega)}^2 \leq C_9\|\bar{\mu}\|_{L_2(\Gamma_1)},$$

where $C_9 = \sqrt{C_2^2 + C_8^2}/2$. Closing in a norm L_2, we shall get the estimate

$$\|\nabla\psi\|_{L_2(\Omega)} \leq C_9\|\mu\|_{L_2(\Gamma_1)}.$$

Then, doing in the same way as in Theorem 2, in virtue of Theorem 1 we obtain

$$\|\psi\|_{L_2(\Gamma_2)} \leq C_{10}\|\mu\|_{L_2(\Gamma_1)},$$

where $C_{10} = C_2(C_9 + C_8)$.

Remark 2. *From Theorem 4 it follows that the conjugate operator to the operator A has the form*

$$A^* : \left\{\mu\middle|\ \mu \in L_2(\Gamma_1),\ \int_{\Gamma_1}\mu ds = 0\right\} \mapsto \left\{h\middle|\ h \in L_2(\Gamma_2),\ \int_{\Gamma_2}hds = 0\right\}$$

and $\|A^*\| \leq C_{10}$.

5 The steepest descent method

The objective functional $J(q) = \|A(q) - f\|_{L_2(\Gamma_1)}^2$ corresponds to the operator equation

$$A(q) = f, \tag{39}$$

whose infimum is obtained under the solution of this equation. To minimize the objective functional the gradient descent method is exploited according to which the sequence $\{q_k\}$ is determined according to the rule

$$q_{k+1} = q_k - \alpha_k J'(q_k), \quad \alpha_k > 0, \quad k = 0, 1, 2..., \tag{40}$$

where q_0 is an initial approach and $\int_{\Gamma_2} q_0 ds = 0$. The parameter α_k is determined using the condition

$$J\left(q_k - \alpha_k J'\left(q_k\right)\right) = \inf_{\alpha>0} J\left(q_k - \alpha J'\left(q_k\right)\right). \tag{41}$$

Calculate $J'(q)$. To do this find the functional increment

$$J\left(q + \delta q\right) - J\left(q\right) = \|A(q + \delta q) - f\|^2_{L_2(\Gamma_1)} - \|A(q) - f\|^2_{L_2(\Gamma_1)} =$$

$$= \int_{\Gamma_1} \left[\left(u\left(x,y,q + \delta q\right) - f\right)^2 - \left(u\left(x,y,q\right) - f\right)^2\right] ds = \tag{42}$$

$$= \int_{\Gamma_1} 2(u(x,y,q) - f)\delta u\, ds + \int_{\Gamma_1} (\delta u)^2 ds,$$

where $\delta u = u\left(x,y,q + \delta q\right) - u\left(x,y,q\right)$, and $u(x,y,q)$ is the direct problem solution corresponding to q.

It is easy to see that δu satisfies the following problem

$$\Delta \delta u = 0, \quad (x,y) \in \Omega, \tag{43}$$

$$\left.\frac{\partial \delta u}{\partial n}\right|_{\partial\Omega} = \begin{cases} 0, & (x,y) \in \Gamma_1, \\ \delta q, & (x,y) \in \Gamma_2. \end{cases} \tag{44}$$

Then for the solution of the problem (43), (44) Theorem 2 is true and the estimate of the type (23)

$$\|\delta u\|_{L_2(\Gamma_1)} \le C_5 \|\delta q\|_{L_2(\Gamma_2)} \tag{45}$$

is fulfilled.

Now using (42) in virtue of (45), the equality

$$J\left(q + \delta q\right) - J\left(q\right) = \int_{\Gamma_1} 2\left(u\left(x,y,q\right) - f\right)\delta u\, ds + O\left(\|\delta q\|^2_{L_2(\Gamma_1)}\right) \tag{46}$$

will be obtained.

Let ψ is the Neumann problem solution (26), (29), where

$$\mu\left(x,y\right) = 2[A(q) - f] = 2\left[u(x,y,q)|_{\Gamma_1} - f(x,y)|_{\Gamma_1}\right], \quad (x,y) \in \Gamma_1. \tag{47}$$

Multiply the equation (43) by ψ and integrate in the domain Ω, using Green's formula and the condition (44)

$$0 = \int_{\Omega} \psi \Delta \delta u\, dx dy = \int_{\Omega} \delta u \Delta \psi\, dx dy + \int_{\partial\Omega} \left(\psi \frac{\partial \delta u}{\partial n} - \delta u \frac{\partial \psi}{\partial n}\right) ds =$$

$$= \int_{\Gamma_1} \delta u \frac{\partial \psi}{\partial n} ds + \int_{\Gamma_2} \psi \delta q\, ds = -\int_{\Gamma_1} 2\left(u\left(x,y,q\right) - f\right)\delta u\, ds + \int_{\Gamma_2} \psi \delta q\, ds.$$

Hence

$$\int_{\Gamma_1} 2\left(u\left(x,y,q\right)-f\right)\delta u ds = \int_{\Gamma_2} \psi \delta q ds. \tag{48}$$

Then using (46), taking into account the equality (48), we have

$$J\left(q+\delta q\right)-J\left(q\right) = \int_{\Gamma_1}\psi\delta q ds + O\left(\|\delta q\|^2_{L_2(\Gamma_1)}\right) \quad .$$

On the other hand

$$J\left(q+\delta q\right)-J\left(q\right) =<\delta q, J'(q)>_{L_2(\Gamma_1)} +O\left(\|\delta q\|^2_{L_2(\Gamma_1)}\right),$$

as a result we obtain that the functional derivative has the form $J'(q) = \psi|_{\Gamma_2}$. And as $A^*\mu = \psi|_{\Gamma_2}$, where μ is defined using (47), then $J'(q) = 2A^*(A(q)-f)$.

Lemma 1. *Let* q, $\delta q \in L_2\left(\Gamma_2\right)$, *then the inequality*

$$|J\left(q+\delta q\right)-J\left(q\right)- <\delta q, J'\left(q\right) >| \le C_5^2 \|\delta q\|^2_{L_2(\Gamma_2)} \tag{49}$$

is fulfilled.

Proof. In fact, in virtue of Remark 1 we have

$$\begin{aligned}
J\left(q+\delta q\right)-J(q) - &<\delta q, J'(q)> = \\
= \|A(q+\delta q)-f\|^2_{L_2(\Gamma_1)} &- \|A(q)-f\|^2_{L_2(\Gamma_1)}- \\
- <\delta q, 2A^*(A(q)-f)> &= \|A(\delta q)\|^2_{L_2(\Gamma_1)} + 2 < A(q)- \\
-f, A(\delta q)> &- 2 < A(\delta q), A(q)-f> = \\
= \|A(\delta q)\|^2_{L_2(\Gamma_1)} &\le \|A\|^2\cdot\|\delta q\|^2_{L_2(\Gamma_2)} \le C_5^2 \|\delta q\|^2_{L_2(\Gamma_2)}.
\end{aligned}$$

Denote

$$\begin{aligned}
f_k\left(\alpha\right) = J\left(q_k - \alpha J'\left(q_k\right)\right) = \\
= \|Aq_k - f\|^2_{L_2(\Gamma_2)} - \alpha\|J'\left(q_k\right)\|^2_{L_2(\Gamma_2)} + \alpha^2\|A\left(J'\left(q_k\right)\right)\|^2_{L_2(\Gamma_2)}.
\end{aligned}$$

Then the minimum of $f_k\left(\alpha\right)$ is obtained for

$$\alpha_k = \frac{\|J'\left(q_k\right)\|^2_{L_2(\Gamma_2)}}{2\|A\left(J'\left(q_k\right)\right)\|_{L_2(\Gamma_1)}} = \frac{\|A^*\left(A\left(q_k\right)-f\right)\|^2_{L_2(\Gamma_2)}}{2\|AA^*\left(A\left(q_k\right)-f\right)\|^2_{L_2(\Gamma_1)}}. \tag{50}$$

State a Lemma whose proof is analogous to that of [3].

Lemma 2. *Let* $f \in L_2\left(\Gamma_1\right)$ *and the solution* $q_* \in L_2(\Gamma_2)$, $\int_{\Gamma_2} q_* ds = 0$ *of the inverse problem* $A\left(q\right) = f$ *exists. Then for the sequential approaches*

$$q_{k+1} = q_k - \alpha_k J'\left(q_k\right) \tag{51}$$

the inequality

$$\|q_{k+1}-q_*\|_{L_2(\Gamma_2)} < \|q_k - q_*\|_{L_2(\Gamma_2)}$$

is fulfilled.

Theorem 5. *(Convergence of the steepest descent method).*

Let $f \in L_2(\Gamma_1)$ and the solution $q_* \in L_2(\Gamma_2)$, $\int_{\Gamma_2} q_* ds = 0$ of the in-
verse problem $A(q) = f$ exists, then if the initial approach q_0 is such that
$\|q_0 - q_*\|_{L_2(\Gamma_2)} \leq M$ and $\int_{\Gamma_2} q_0 ds = 0$, then the approaches q_k converge in
functional, and the estimate

$$I(q_k) \leq \frac{C_5^2 M^2}{k}, \quad k = 1, 2, ...$$

is fulfilled.

Proof. Note, as $\int_{\Gamma_2} q_0 ds = 0$, and from $\int_{\Gamma_2} \psi ds = 0$ it follows that
$\int_{\Gamma_2} J'(q_0) ds = 0$, then $\int_{\Gamma_2} q_k ds = 0$, $k = 1, 2...$. In (49) take $q = q_k$, $\delta q = -\alpha J'(q_k)$, where $\alpha > 0$, then we will have

$$\left| J(q_k - \alpha J'(q_k)) - J(q_k) + \alpha \|J'(q_k)\|_{l_2(\Gamma_2)}^2 \right| \leq C_5^2 \alpha^2 \|J'(q_k)\|_{L_2(\Gamma_2)}^2.$$

In turn, from (40) and (41) we have

$$J(q_{k+1}) = J(q_k - \alpha_k J'(q_k)) \leq J(q_k - \alpha J'(q_k)).$$

Then from the last two estimates

$$\alpha(1 - C_5^2\alpha)\|J'(q_k)\|_{L_2(\Gamma_2)}^2 < J(q_k) - J(q_{k+1}) \tag{52}$$

follows.

The inequality (52) is also fulfilled for the maximum value of $\alpha = \frac{1}{2C_4^2}$

$$\frac{1}{4C_5^2}\|J'(q_k)\|_{L_2(\Gamma_2)}^2 < J(q_k) - J(q_{k+1}). \tag{53}$$

According to the condition of the theorem $A(q^*) = f$ is fulfilled. Then

$$J(q) = < Aq - f, Aq - f > = < q - q_*, A^*(A(q) - f) > =$$

$$= \frac{1}{2} < q - q_*, 2A^*(A(q) - f) > = \frac{1}{2} < q - q_*, J'(q) > .$$

Hence, in virtue of the Cauchy inequality

$$J(q) \leq \frac{1}{2}\|q - q_*\|_{L_2(\Gamma_2)} \cdot \|J'(q)\|_{L_2(\Gamma_2)} \tag{54}$$

from (5.16) we have

$$\frac{1}{4}\|J'(q)\|_{L_2(\Gamma_2)} \geq \frac{J^2(q)}{\|q - q_*\|_{L_2(\Gamma_2)}}, \tag{55}$$

since $q \neq q_*$.

According to the requirement of the theorem $\|q_0 - q_*\|_{L_2(\Gamma_2)} \leq M$, from Lemma 2

$$\|q_k - q_*\| \leq M \tag{56}$$

follows.

Then from (55) in virtue of (56) and (56) we will get

$$\frac{J^2(q_k)}{M^2} \leq \frac{J^2(q_k)}{\|q_k - q_*\|_{L_2(\Gamma_2)}^2} \leq \frac{1}{4}\|J'(q_k)\|_{L_2(\Gamma_2)}^2 \leq C_5^2(J(q_k) - J(q_{k+1})). \tag{57}$$

Rewrite (5.19) denoting $a_k = J(q_k)$ and dividing both parts of the inequality by C_5^2, as a result we shall get

$$\frac{a_k^2}{C_5^2 M^2} \leq a_k - a_{k+1}. \tag{58}$$

Now in virtue [6] from (5.20) the estimate

$$a_k \leq \frac{C_5^2 M^2}{k}, \quad k = 1, 2, \ldots$$

follows, it means that

$$J(q_k) \leq \frac{C_5^2 M^2}{k}, \quad k = 1, 2, \ldots .$$

References

1. Kabanikhin SI, Karchevskiy AL (1995) J Inv Ill-Posed Problems 3(1):21–46
2. Kabanikhin SI, Bektemessov MA, Ayapbergenova AT, Nechayev DV (2004) Comput Technol 9(49–60) (in Russian)
3. Kabanikhin SI, Bektemessov MA, Nechayev DV (2003) Inverse Probl Eng Mech 4: 447-456
4. Ladyzhenskaya OA, Uraltseva NN (1973) Linear quasi-linear equations of elliptic type. Nauka, Moskow (in Russian)
5. Sobolev SL (1988) Some applications of functional analysis in mathematical physics. Nauka, Moskow (in Russian)
6. Vassilyev FP (1980) Numerical methods of solving extreme problems. Nauka, Moskow (in Russian)

Challenges of future hardware development and consequences for numerical algorithms

U. Küster

High Performance Computing Center Stuttgart (HLRS), University of Stuttgart, Nobelstraße 19, 70569 Stuttgart, Germany resch@hlrs.de

Summary. We give an overview of some contemporary processor developments and draw conclusions for developments of numerical algorithms. Multicore CPUs are developed by nearly all hardware manufactures. The bandwidth peak performance relation is decreasing. Both have consequences for the efficiency of todays computer codes. We propose a Finite Difference discretization technique which is decreasing the data intensity.

Introduction

The development of modern processors in the last decade was subdued to the idea of increasing the density of the active elements on the chip. This idea became a self understanding principle of 'Moores Law' stating a density doubling every 18 – 24 months. The frequency increase was a consequence to density increase. People got faster processors and larger caches at the same time. Changes in architecture were moderate. As economically most important the IA 32 processor line was never forced to change the instruction set substantially because of the fast development.

Some well foreseen difficulties became clearer in the last years. The increase of the frequency results in the increase of the relative memory and cache latencies. But also the power dissipation became larger and larger. Even if active cooling may handle 300 Watt per chip it seems implausible that people will accept desktops with power consumption of an electric iron. This leads to stagnating frequencies and stagnating serial performance of the processors of the near future. Performance improvement will be due only to parallel capabilities. The higher integration gives the opportunity of implementing additional functional units or additional processors on the same chip. The aggregated (contrary to serial) floating point performance of a chip will raise also in future. But a severe bottleneck may be the aggregated memory bandwidth. This will lead to an incredible floating point performance potential limited for usage for small data sets.

Computer architects offer solutions for problems with a still unknown benefit. How to react to these offers from the algorithms side? In the following we give some annotations about modern chips and about implications for numerical algorithms.

1 Contemporary standard hardware development

1.1 Modern standard processors

Smaller feature sizes enable larger caches and more processor cores per chip. Even recognizing that the gate length today is to be measured in atom numbers we will see chips manufactured in 65 nm technology. That means that half the distance between two interconnecting wires is 130 atoms. Remark that the distance of two silicon atoms in a crystal is 0.5 nm . The frequency will remain constant to limit the loss of electrical power. What will be done with the high number of active elements on the chip? First is to increase the on chip cache sizes. We see in table 1 the remarkable number of 1.7 Bill transistors for the Intel IA-64 Montecito. They are mainly used for two large L3 caches of 12 MB each. They take roughly $\frac{2}{3}$ of the chip size. But there are also two cores on the same chip. Intel Montecito as well as the Intel Dempsy, AMD Opteron, IBM Power 4 and 5, Sun UltraSparc IV+ have two cores on chip. Four cores will be seen in the next years. The new SUN Niagara concept is still more radical. Eight UltraSparc III cores with 4 hardware threads each will be implemented. They will have a low frequency and a low power consumption of less than 60 Watt. The caches are relatively small. Latency hiding is the target of multithreading architectures like Niagara. As long as one thread is waiting for data another thread may be active on the functional units. This architecture assumes sufficient waiting threads in a throughput environment. The Niagara processor is not made for numerical applications. Its floating point capabilities are very small. But assuming a chip with 32 independent hardware threads we find the numerical problem to have a lot of inherent parallelism. Helpful are small on chip latencies and the remarkable bandwidth. Performance will be aggregated and not serial. Remember the Tera MTA machine which handled 128 hardware threads on one processor.

SUN Niagara paves the way for other chip developments to follow. For AMD and the two main Intel lines we see four and more cores on chip to come. But despite increasing bandwidth the per core available bandwidth will be the bottleneck. Table 1 gives an overview of some actual and expected processors in comparison of die size, number of transistors, technology, frequency and bus bandwidth.

We have omitted the proprietary developments of CRAY, SGI, NEC and HP. SGI and HP are using the Intel IA 64 processors. HP is still developing the PA-RISC line. CRAY will mix diverse processors in the same machine (Rainier concept). Their Black Widow vector processor features high memory bandwidth as for the future NEC processor. Multi core chips are non disclosed.

Table 1. Properties of some contemporary and future processors

Name	die size (mm^2)	no trans (Mio)	technology (nm)	cores	frequency (GHz)	BW (GB/s)
AMD Opteron	200	230	90	2	2.4	6.4
SIT Cell Processor	221	230	90	1+8	4.0	25.6
IBM Power 5	389	?	130	2	2.0	16
Intel Dempsy	?	?	65	2	3.8	8.5
Intel IA-64 Madison	432	592	130	1	1.6	6.4
Intel IA-64 Montecito	580	1720	90	2	2.0	10.6
SUN US IV+	?	295	90	2	2.0	2.4
SUN Niagara	?	?	90	8	1.0 – 1.5	20

1.2 Clearspeed accelerator card CSX 600

The Clearspead accelerator card CSX 600 ([1]) consists of two processors with their own memory on a PCI-X card. Cards for other interfaces are to be expected. The Clearspeed processor runs at a low frequency of 250 MHz. Thus the power consumption could be reduced to only 5 Watts per processor. The chip is composed of 96 processing elements (PE) which are conducted by a small host. Each PE has its own fast SRAM memory of (only) 6 KB. The internal bandwidth of 1 GB/s per PE is high enough to feed the functional units of 0.25 GFLOPs. But the external memory bandwidth of 3.2 GB/s is by far to low to pump the data into the processing elements. By this reason there are only a few exceptions proving the aggregated performance of 50 GFLOPs for the two processor cards. Also DGEMM matrix multiplication is not able to get more than half the peak performance. But even 25 GFLOPs are excellent for some special applications as dense matrix operations, fast fourier transforms and some techniques of molecular dynamics. Loading the data from the host computer to the accelerator card may be time consuming at least for solving a set of small problems. This will be better for future interfaces.

The parallel programming model is based on a C-dialect with additional variable declaration syntax. Special data movement instructions allow for distributing data to the processing elements. The model emphasizes on running copies of identical procedures with different data (Flynn SPMD).

Clearspeed shows that high aggregated performance can be reached by small frequency and low power consumption. But we notice an insufficient memory bandwidth limiting the sustained performance for problems with modest data intensity.

1.3 The IBM Sony Toshiba Cell Processor

The development of Cell Processor is done in a collaboration of IBM, Sony and Toshiba ([2, 3, 4]). The Cell Processor will serve for entertainment electronics. It will empower the future Sony Playstation 3 for high performance graphics and will be used for computing intensive High Definition Television (HDTV). The processor may be used in medical applications and signal processing. Workstation systems for scientific usage can be expected. The Cell Processor will be a cheap device despite his complexity.

The machine has an uncommon architecture. It is a single chip architecture with a PowerPC processor as host. This processor will run the operating system and control the different tasks. The processor allows for two hardware threads and operates in order. The Altivec/VMX vector part allows for relatively high performance. Instruction and data L1-caches have 32 KB. The 512 KB L2-cache is on die. The system has two XDR Rambus memory interfaces with high memory bandwidth of 25.6 GB/s.

Connected to the system caches is the Element Interconnect Bus (EIB). This bus consists off 2 rings in both directions operating at half of the system clock. The peak bandwidth is 96B/cycle. For a 4 GHz system this results in 192 GB/s. The EIB feeds eight Synergistic Processor Elements (SPE). These SPEs are independent vector processors. Each has its own local memory of 256 KB. A register file with 128 registers of 128 bit width holds the vector operands. Two instructions are issued per clock cycle and operated in(!) order. Four single precision floating point multiply-add instructions may be handled at the same time. The single precision peak performance of all SPEs of a 4 GHz processor is around 250 GFLOPs! The single precision units deliver unrounded results restricting their usability to graphical output. The IEEE double precision performance of 26 GFLOPs is by far less but still four times more than todays PCs peak performance. The memory bandwidth peak performance relation of 1 B/FLOP is a reasonable value for double precision results. For single precision results this relation is quite small with 0.1 B/FLOP. A special DMA engine in the SPE maps parts of the main memory to the local memory and initiates and performs block data transfers. The block data transfers may also be performed between different SPEs of different cells.

Different cells are connected by the fast FLEXIO interface of aggregated 76 GB/s(!) bandwidth.

The one chip system has an incredible peak performance. With respect to numerical applications cutting of floating point results has to be considered as deficiency. But with the actual IEEE double precision performance a very challenging system has to be acknowledged. Raising double precision capabilities can be expected in future systems.

Noticing the fast memory system it is important to keep an eye on the decreasing bandwidth peak performance relation. The architects try to smooth the consequences by the SPEs local memories. They enable the programmers to optimize the codes by decoupling heavily used data from main memory. But the programmer is also forced in doing that.

Questionable is the ability of porting important numerical algorithms to this architecture. We have a set of high performance processors with unusual memory access. Data have to be transferred actively.

2 Requirements for numerical software

Because of the deviant gradient of the agglomerated chip performance and the system memory bandwidth programmers have to be aware, that the amount of data needed for floating point operations has to be decreased as long as they want to participate in the flow of performance increase. We see in table 2 the memory intensity of some well know algorithms. We assume double

Table 2. Data intensity of some algorithms and their blocking potential

algorithm	Byte / Flop	blocking potential
Jakobi iteration for Poisson eq	9.14	weak
multigrid for Poisson eq	6.31	low
CG with sparse matrix	14.6	weak
Roe flux splitting CFD	0.44	good
sparse matrix vector multiply	14	weak
Eispack SVD (non blocked)	10	high for small problems
Lapack SVD (blocked)	10	good
Lattice Boltzmann NS (BEST)	1.77 - 1.87	weak
dense matrix oper Multiply solution of lin sys	$O(1/dim)$	high

precision arithmetic and do not take into account that data could be reused in a temporal or spatial context. Some parts of the algorithms may have higher data intensity. Numerical problems are frequently as large as the system memory allows. This remark is important in consideration of fixed sized benchmarks like SPEC FP serving as guidelines for computer architects. For iterative schemes using a large number of outer loop iterations that means the periodical reading of a relevant part of the memory, transforming and writing back a large fraction of memory. This applies for explicit schemes, Krylov space algorithms, multigrid algorithms. Fixing a small data set for a longer time is not possible except for techniques involving nontrivial operations on dense matrices. Even for small (dimension > 10) dense matrices the blocking potential is high for multiplication, inversion, solution of linear systems, eigenvalue problems (not for addition).

3 Finite Differences revised

Beyond **Finite Element** and **Finite Volume** the **Finite Difference** discretization is a very general approach for solving partial differential equations. Whereas the first is coming from the handling elliptic and parabolic equations but in between also used for solving hyperbolic and convection dominated problems by special Galerkin techniques the second is well suited in solving conservative equations by using directly the flux formulation of the Gauss theorem. Both deliver also weak solutions of the partial differential equations. The

Finite Difference formulation allows for very general equations but typically not for weak solutions. Finite difference schemes are mostly used on orthogonal equidistant grids or regularly structured grids as transformed equidistant grids. The metric coefficients of the transformation are calculated by Finite Differences themselves. The partial differential equations have to be replaced by their transformed counterparts. As long as nonlinear equations are treated, linearized equations are solved.

In the **Finite Element** context this means that the small finite element matrices have to be recalculated again and again and assembled into the stiffness matrix. The sparsity structure of this matrix is conserved. Only the values are to be changed. The amount of work due to the (nonlinear) element matrices is high especially for complicated elements. Handling these is perfectly parallelizable. They even meet the condition of having a lot of operations for only a few data (node coordinates und state variables). These requirements meet perfectly a multiple core environment with limited bandwidth. The matrix assembly phase has very different requirements. Only a few additions are to be applied for all the matrix element data. The situation is even worse because nodes have contributions of several elements. Parallelization forces in using coloring techniques or very fast locking of individual nodes. Each color allows for the parallel assembly of a part of the matrix. A small amount of operations on dependent data for the large matrix data set are contrary to a multicore system with limited bandwidth. The next step is the solution of the resulting large linear system. Direct solvers will work only for limited problemsizes otherwise they need to much computing time and memory. They work in parallel (multifrontal approach) but only for a limited number of processors. Iterative procedures as Krylov space procedures involve the sparse matrix vector multiplication as time determining part. Flexible sparse matrix data structures involve heavily indirect addressing and high data intensity in contrast to the limited processor bandwidth. This operation may be well parallelized by domain decomposition but at the expense of frequent communication steps. Tightly connected on chip cores with a communication distance of only some cycles are useful as long as the processor locality in the shared memory system may be controlled by the user. Preconditioners have to have a non recursive parallelizable structure. Incomplete LU factorization will only work on domain patches.

Finite Volume techniques omit typically the matrix as data object. Otherwise the same remarks as for the Finite Element case have to be applied. Important part of finite volume schemes is the calculation of fluxes across the cell faces. This calculation step involves a lot of operations for a relatively small set of data. But the amount of data is filling a large fraction of the memory. This step is also perfectly parallelizable by domain decomposition. Adding up the fluxes to the cell status value involves only a some few operations for the large amount of flux data. Blocking can help to reduce the amount of memory traffic needed. Communication and synchronization and dependent on the algorithm coloring is needed.

Also for **Finite Differences** we have obligatory need of data intensive matrix schemes. Parallelization is well experienced but leads to high bandwidth needs in shared memory systems. A good idea would be the decomposition of the large sparse matrix in a smaller set of dense block matrices under the condition that the numerical efficiency is not worse. In this case the performance will be increased by exploiting the faster dense matrix vector multiplication which has smaller data intensity. Also preconditioning of the matrix may be set up on top of these blocks. Parallelization must not divide these blocks to simplify the approach. The matrix blocks and the conforming vectors could be build by combining node states or node states of adjacent nodes. But this approach may end in unnecessary operations.

Any processor will be engaged by a sequence of (small) dense matrix vector operations. The matrices should be large enough so that the processors computing times exceed significantly the time for harmful synchronizations or atomic operations.

In the following we redefine Finite Difference discretization in a way leading immediately to small dense matrices and vectors. We propose a mechanism which can be understood as generalization of the Finite Difference approach for the discretization of linear partial differential equations. This mechanism combines compact stencils or node neighbourhoods with high order discretizations. It shows the way to expand node local solutions in polynomials and other functions. It works in regular and unstructured grids; more precisely, it works in arbitrary node neighbourhoods as long as the neighbourhood is geometrically rich enough. The mechanism forces to solve small dense matrix problems to define the discretization weights for each node. The solver for the linear global system involves small dense block matrices in a simple way. The nodewise effort is substantially larger than for conventional Finite Difference techniques. But the overall numerical effort for reaching the same accuracy should be smaller due to a better approximation order. The performance is essentially higher for appropriate implementations. The technique is including conventional Finite Difference discretization for arbitrary linear operators as special case.

We assume Ω to be a region in an $1 - 4$ dimensional space. The scalar linear inhomogeneous partial differential equation

$$L(\Phi(\hat{x})) = r(\hat{x}), \quad \forall \hat{x} \in \Omega, \tag{1}$$

is to be solved. Most of the next remarks can be generalized to partial differential equation systems. Assuming the node $\hat{x}_j \in \Omega$ for some $j \in \mathcal{N}$ the equation will be discretized by

$$\sum_{k \in \mathcal{N}_j} \alpha_j^k \Phi_k = R_j, \quad \forall j \tag{2}$$

$\mathcal{N}_j \subset \mathcal{N}$, is a small discrete neighbourhood of the node $\hat{x}_j$. The weighting coefficients α_j^k are $q \times q$-matrices instead of scalars in the usual case. The values

Φ_k are q-dimensional vectors. They serve as coefficients for a set of local approximation functions. These functions could be some low order polynomials. Also special functions can be used.

This leads to the question of determine the coefficient matrices in a consistent way. At any node $\hat{x}_j$ we take a matrix of approximation functions

$$\Omega \ni \hat{x} \longrightarrow S_j(\hat{x}) \in \mathcal{R}^{q \times r}. \tag{3}$$

The number r of equations to be fulfilled will typically exceed the block size q. A linear combination of these functions will provide for an approximation to the local solution. We postulate the discretization of these approximation functions to be exact. This results in the following requirements for the coefficient matrices α_j^k.

$$\sum_{k \in \mathcal{N}_j} \alpha_j^k S_j(\hat{x}_k) = \left[\alpha_j^{k_1}, \cdots, \alpha_j^{k_{|\mathcal{N}_j|}} \right] \begin{bmatrix} S_j(\hat{x}_{k_1}) \\ \cdots \\ S_j(\hat{x}_{k_{|\mathcal{N}_j|}}) \end{bmatrix} = L(S_j(.)|_{\hat{x}_j}). \tag{4}$$

The left hand side is the difference operator at node $\hat{x}_j$ and the right hand side simply the differential operator applied to the matrix of the approximation functions. We get q linear systems with a $(q * |\mathcal{N}_j|) \times r$ matrix. The system can be solved in a nonunique way as long as $(q * |\mathcal{N}_j|) \geq r$ and the matrix has no rank defect. The solution ensures that the discretization has an order at least as high as the order of the used approximation functions after substraction of the order of the differential operator. Helpful symmetries of neighbourhoods are neglected. This is to be applied for the polynomial case. For a non polynomial case we have to ensure that the approximation functions are able to approximate polynomials up to a certain order. The linear coefficient equations have to be solved at each node. For higher dimensional cases (3-D, 4-D) this will be a significant effort. For grids with identical geometrical neighbourhoods the system has to be solved only once. We remark that the rows of the coefficient matrices $\left[\alpha_j^{k_1}, \cdots, \alpha_j^{k_{|\mathcal{N}_j|}} \right]$ are subjected to the same linear system with different right hand sides. If the degrees of freedom $q * |\mathcal{N}_j|$ are sufficiently large 'upwinding' for better stability of the overall system can be reached. The system may be rewritten in a way that the coefficient matrix of the reference node $\hat{x}_j$ (assumed to be the identity) and the former right hand side change their roles. This ensures that the diagonal blocks of the system matrix are identity matrices advantageous for iterative procedures. The coefficient matrices have to be calculated and stored for all nodes. They dependent on the local neighbourhood. The calculation involves only a few defining data but a large number of operations. It can be done for all nodes in parallel without any interference. One way for the calculation is the Singular Value Decomposition.

An obvious example for the approximation matrices are polynomials with all monomials up to a certain order and all their derivatives with respect to the cartesian space directions up to a smaller order. The following example $(\hat{x}_j = 0)$ shows a polynomial matrix with order 4 and derivative order 2. The sequence of the derivatives is the same as the sequence of the monomials. We see that the matrix may be decomposed in two parts. The first part $S_1(x)$ ($x = (x_1, x_2)$) is invertible and has the algebraic property $S_1(x+y) = S_1(x)\,S_1(y)$ for all x, y.

$$S(x) = [S_1(x), S_2(x)] =$$

	C1	$\frac{x_1^1}{1!}$	$\frac{x_2^1}{1!}$	$\frac{x_1^2}{2!}$	$\frac{x_1^1 x_2^1}{1!\,1!}$	$\frac{x_2^2}{2!}$	$\frac{x_1^3}{3!}$	$\frac{x_1^2 x_2^1}{2!\,1!}$	$\frac{x_1^1 x_2^2}{1!\,2!}$	$\frac{x_2^3}{3!}$	$\frac{x_1^4}{4!}$	$\frac{x_1^3 x_2^1}{3!\,1!}$	$\frac{x_1^2 x_2^2}{2!\,2!}$	$\frac{x_1^1 x_2^3}{1!\,3!}$	$\frac{x_2^4}{4!}$
∂^0	1	$\frac{x_1^1}{1!}$	$\frac{x_2^1}{1!}$	$\frac{x_1^2}{2!}$	$\frac{x_1^1 x_2^1}{1!\,1!}$	$\frac{x_2^2}{2!}$	$\frac{x_1^3}{3!}$	$\frac{x_1^2 x_2^1}{2!\,1!}$	$\frac{x_1^1 x_2^2}{1!\,2!}$	$\frac{x_2^3}{3!}$	$\frac{x_1^4}{4!}$	$\frac{x_1^3 x_2^1}{3!\,1!}$	$\frac{x_1^2 x_2^2}{2!\,2!}$	$\frac{x_1^1 x_2^3}{1!\,3!}$	$\frac{x_2^4}{4!}$
∂_{x_1}		1	0	$\frac{x_1^1}{1!}$	$\frac{x_2^1}{1!}$	0	$\frac{x_1^2}{2!}$	$\frac{x_1^1 x_2^1}{1!\,1!}$	$\frac{x_2^2}{2!}$	0	$\frac{x_1^3}{3!}$	$\frac{x_1^2 x_2^1}{2!\,1!}$	$\frac{x_1^1 x_2^2}{1!\,2!}$	$\frac{x_2^3}{3!}$	0
∂_{x_2}		0	1	0	$\frac{x_1^1}{1!}$	$\frac{x_2^1}{1!}$	0	$\frac{x_1^2}{2!}$	$\frac{x_1^1 x_2^1}{1!\,1!}$	$\frac{x_2^2}{2!}$	0	$\frac{x_1^3}{3!}$	$\frac{x_1^2 x_2^1}{2!\,1!}$	$\frac{x_1^1 x_2^2}{1!\,2!}$	$\frac{x_2^3}{3!}$
$\partial_{x_1}^2$				1	0	0	$\frac{x_1^1}{1!}$	$\frac{x_2^1}{1!}$	0	0	$\frac{x_1^2}{2!}$	$\frac{x_1^1 x_2^1}{1!\,1!}$	$\frac{x_2^2}{2!}$	0	0
$\partial_{x_1 x_2}$					1	0	0	$\frac{x_1^1}{1!}$	$\frac{x_2^1}{1!}$	0	0	$\frac{x_1^2}{2!}$	$\frac{x_1^1 x_2^1}{1!\,1!}$	$\frac{x_2^2}{2!}$	0
$\partial_{x_2}^2$						1	0	0	$\frac{x_1^1}{1!}$	$\frac{x_2^1}{1!}$	0	0	$\frac{x_1^2}{2!}$	$\frac{x_1^1 x_2^1}{1!\,1!}$	$\frac{x_2^2}{2!}$

For this example we have $q = 1 + 2 + 3 = 6$ and $r = 1 + 2 + 3 + 4 + 5 = 15$ in equation 3.

We have still to clarify how to get the necessary additional equations. We claim that the first line of the approximation matrix will comply with the right hand side of the partial differential equation

$$e_1^T L(S_j(.)|_{\hat{x}} \Phi_j) = r(\hat{x}) \tag{5}$$

in the neighbourhood $\mathcal{N}_j$ of the node $\hat{x}_j$ for the solution vector Φ_j. To get this we have to determine the right hand side vector R_j in equation 2 . This can be done in the following discretized way. We determine a (nonunique) parameter vector a_r fulfilling the equations

$$e_1^T L(S_j(.)|_{\hat{x}_k} a_r) = r(\hat{x}_k), \quad \forall k \in \mathcal{N}_j. \tag{6}$$

Then the right hand side vector will defined by

$$R_j = L(S_j(.)|_{\hat{x}_j} a_r). \tag{7}$$

This is the way in getting the additional equations. In the context of the partial differential equation we claim its validity also for its derivatives (different to conventional discretization). The approach for determination of R_j is independent on the approximation functions S_j.

Handling the boundary conditions becomes more difficult as for conventional discretization. It is similar to the specification of the right hand side but is involving the geometry of the boundary. We analyse here only the 2D case and only Dirichlet boundary conditions. We claim that for a parametrization of the boundary $\partial \Omega$ the approximation functions

$$r \mapsto S(\hat{x}(r)) \in \partial \Omega \tag{8}$$

are discretized exactly. Then it can be shown that the following system

$$D_{Rand}(\hat{x}(r)) = \left[\begin{array}{c|cc|ccc} 1 & & & & & \\ 0 & \partial_r x_1 & \partial_r x_2 & & & \\ 0 & \partial_{rr} x_1 & \partial_{rr} x_2 & \partial_r x_1^2 & 2\partial_r x_1 \partial_r x_2 & \partial_r x_2^2 \end{array} \right], \tag{9}$$

$$\sum_{k \in \mathcal{N}_j} \alpha_j^k S(\hat{x}_k) = D_{Rand} S(\hat{x}(r)_j) \tag{10}$$

determines the coefficient matrices α_j^k for a boundary node $\hat{x}_j$. Here we have considered only terms of second order.

With these coefficient matrices the linear equation

$$\sum_{k \in \mathcal{N}_j} \alpha_j^k \Phi_k = \left[\begin{array}{c} f(\hat{x}(r)) \\ \partial_r f(\hat{x}_j) \\ \partial_{rr} f(\hat{x}_j) \end{array} \right] \tag{11}$$

is to be solved on the boundary on all boundary nodes. For this purpose we also need the tangential derivatives of the boundary condition f on the boundary. If this procedure is not appropriate a similar approach as defined for the right hand side may be used.

If the global linear system has been solved for the vectors Φ_j then new parameter vectors Ψ_j are defined by the equation

$$\Phi_j = S_j(\hat{x}_j) \Psi_j \tag{12}$$

They deliver the function

$$\hat{x} \longrightarrow S_j(\hat{x}) \Psi_j. \tag{13}$$

as the local solution. The first part of

$$S_j(\hat{x}) = \left[S_j^1(\hat{x}), S_j^2(\hat{x}) \right] \tag{14}$$

results in

$$\Psi_j = S_j^1(\hat{x}_j)^{-1} \Phi_j. \tag{15}$$

For the polynomial approximation matrix S_j we have simply

$$S_j^1(\hat{x}_j) = I. \tag{16}$$

The following tables 3 and 4 show the dependency of the interpolation order, the block sizes q, the number of approximation polynomials r to be exactly discretized and the number of neighbours needed for 2D (table 3) and 3D (table 4). The order of the differential operator has to be substracted from the interpolation order. It can be seen that the interpolation order may be high even for relatively small neighbourhoods.

Table 3. Block sizes, number of test functions, necessary neighbours in 2D

block order	intpol order	block size q	test fun r	min node number
0	1	1	3	3
0	2	1	6	6
0	3	1	10	10
0	4	1	15	15
0	5	1	21	21
1	2	3	6	2
1	3	3	10	4
1	4	3	15	5
1	5	3	21	7
2	3	6	10	2
2	4	6	15	3
2	5	6	21	4
2	6	6	28	5
2	7	6	36	6
2	8	6	45	8
3	5	10	21	3
3	6	10	28	3
3	7	10	36	4
3	8	10	45	5
3	9	10	55	6
3	10	10	66	7

Table 4. Block sizes, number of test functions, necessary neighbours in 3D

block order	intpol order	block size q	test func r	min node number
0	1	1	4	4
0	2	1	10	10
0	3	1	20	20
0	4	1	35	35
0	5	1	56	56
1	2	4	10	3
1	3	4	20	5
1	4	4	35	9
1	5	4	56	14
1	6	4	84	21
1	7	4	120	30
2	3	10	20	2
2	4	10	35	4
2	5	10	56	6
2	6	10	84	9
2	7	10	120	12
2	8	10	165	17
3	5	20	56	3
3	6	20	84	5
3	7	20	120	6
3	8	20	165	9

The global linear system can be solved by a Krylov space procedure. The matrix consists on blocks reducing the needs for indirect addressing. Even direct methods could be used for smaller cases.

The method has not yet been tested.

To collect the properties of the scheme: The costs per node are significantly higher. But the order of the method is also higher. The stencil is not increased. The administrative part and the communication mechanism are not changed compared to the conventional counterpart. The parallelization strategies are not changed. The halos of the domains remain the same. The number of messages is not changed but their size. This fits the trend of computer architecture because communication bandwidth will be slowly increased but latency is going down much slower We have to be aware of some theoretical and practical difficulties. The stability of the global system has to be ensured. Important are the questions of generalizing the mechanism to nonlinear equations and to finite volume schemes.

Conclusion

Numerical researchers have to recognizes that the trend of computer architectures is moving to a worse bandwidth peak performance relation combined with multicore chips. They have to be aware to decrease the relative data consumption. The number of operations done is less important than the amount of memory data needed for that purpose. Differently spoken is it allowed to make floating point operations as long as they need no additional data. We have proposed a Finite Difference mechanism involving small sized dense matrices. The additional effort for this technique may be recompensated by its higher order. Additional investigations have to ensure the benefit of the method.

References

1. http://www.clearspeed.com/products/si.php
2. http://www-306.ibm.com/chips/techlib/techlib.nsf/products/Cell
3. http://www.ibm.com/developerworks/power/cell/
4. http://researchweb.watson.ibm.com/journal/rd/494/kahle.pdf

Simulation of flame propagation in closed vessel with obstacles

A. Kaltayev and Zh. Ualiev

Kazakhstan kaltayev@kazsu.kz

1 Introduction

Among various numerical techniques for partial differential equations, the finite difference method is one of the most attractive. These methods on structured meshes are usually faster than on unstructured meshes and in the case of regular domain it is very efficient in terms of computation cost. In this sense, a fictitious domain method (FDM) allows to work in a regular domain using a regular meshes independently of the geometry of the problem investigated. FDM techniques for partial differential equations have recently shown an interesting potential for solving complicated problems in Science and Engineering [1], [2], [3]. The main reason for the popularity of FDMs is that they allow to use structured meshes on a geometrically simple shape (typically a rectangle in 2D) together with an auxiliary (fictitious) domain containing the actual domain, therefore allowing the use of fast solvers. Another important point is that the stability condition of the resulting scheme is the same as the one of the finite difference scheme [1]. Up to now a FDM was used for solving complicated problems of physics with Dirichlet boundary conditions [4], [5],[6]. But many problems of science (for example such as reacting flow problems or combustion problems) are Neumann problems and here the application of FDM has essential peculiarities. In the context of the study a fictitious domain method is extended and implemented for combustion problems in a non-regular domain. The most prominent example of these problems is the propagation of a flame in a tube with obstacles. It is known that the combustion process of premixed gases in tubes or vessels is strongly affected by obstacles. For freely propagating flames obstacles can cause violent flame acceleration and enhance the transition from deflagration to detonation.

2 Physical, chemical and mathematical models

2.1 Physical and chemical models

We are considering the cases when laminar flame propagates in a closed rectangular tube with length L and cross section of $2h$ with obstacles placed at various distances from the left side wall. We focus our attention on its numerical solution by FDM approach to study the hydrodynamic structure of the flame.

A line igniter with circular cross section, placed at the centerline near the left side wall and perpendicular to the $x - y$ plane, initiates the combustion of the combustible gas mixture. So that the physical model assumes a flow in the tube is planar and symmetric.

The study is based on the assumption of a low Mach number flow (sound wave propagation is neglected) and uses a one-step global chemistry model of the laminar flame. The combustible gas is a stoichiometric methane-air mixture.

Furthermore, the following assumptions are made:

a) the molecular weights of the species are equal, $W_i = W$;
b) the viscosity of the mixture is $\mu = \mu_0 (T/T_0)^{0.75}$, where μ_0, T_0 are the viscosity and temperature of unburned gas at initial time;
c) the specific heat of the mixture is (CRC Handbook, 1982-1983)

$$c_p(T) = c_{pn}(T) \sum_i Y_i e_i,$$

where

$$c_{pn}(T) = (970 + 218 \cdot T/1000 - 32\,(T/1000)^2)\,J/(kg \cdot K)$$

is the nitrogen heat capacity, e_i are constants;
d) the Prandtl number $Pr = \lambda/\mu c_p$ is constant;
e) the binary diffusion coefficients of all pairs of species are equal; and Lewis number $Le = \lambda/\rho D c_p = 1$ is assumed.

A single step, global irreversible reaction

$$CH_4 + 2O_2 + 7.52N_2 \rightarrow CO_2 + 2H_2O + 7.52N_2$$

is used with Arrhenius type reaction rate [7]

$$\frac{d}{dt}[CH_4] = -A[CH_4][O_2]\,exp(-\frac{T_a}{T}) = \dot{w}_{CH_4}/W,$$

where $[\cdot]$ denotes molar concentration, $A = 2.4 \cdot 10^{10}\,m^3/mole \cdot sec$, $T_a = E/R = 24370\,K$. Using these Arrhenius parameters the calculated flame speed u_n and the flame thickness l at the standard conditions are 40 cm/sec and 0.5 mm respectively which is in agreement with experimental data.

2.2 The mathematical model

With the assumptions given above, the governing conservation equations of energy, species, state, mass and momentum in the low Mach number limit can be written as follows [8], [9]:

$$\rho\frac{\partial H}{\partial t} + \mathbf{V} \cdot grad\, H = \frac{dP_t}{dt} + div\left(\frac{\lambda^\epsilon}{c_p}grad\, H\right) + S_{ig}, \tag{1}$$

$$\rho\frac{\partial Y}{\partial t} + \mathbf{V} \cdot grad\, Y = \dot{w}_{CH_4} + div(\rho D^\epsilon\, grad\, Y), \tag{2}$$

$$H = c_p(T)T + QY, \tag{3}$$

$$\frac{dP_t(t)}{dt} = \frac{\kappa_f - 1}{\mathcal{V}}\left(\int_{\mathcal{S}}\lambda\frac{\partial T}{\partial n}d\mathcal{S} + Q\int_{\mathcal{V}}\dot{w}_{CH_4}d\mathcal{V}\right), \quad \mathcal{V} = 2h^2 L. \tag{4}$$

$$P_t(t) = \frac{\rho RT}{W}, \tag{5}$$

$$div\, \mathbf{V} = -\frac{\partial\rho}{\partial t}, \tag{6}$$

$$\frac{\partial \mathbf{V}}{\partial t} + div(\mathbf{v} \otimes \mathbf{V}) = -grad\, \Pi + div(2\mu\dot{S}) - \xi^\epsilon\mathbf{V}, \tag{7}$$

where t denotes the time, ρ - the density, H - the enthalpy, $\mathcal{V}$ and $\mathcal{S}$ - the volume and surface of the vessel, κ_f - the ratio of specific heats within the reaction zone, Y - the mass fraction of methane, Q - the heat of combustion, R - the universal gas constant, $\mathbf{v}$ - the velocity and $\mathbf{V} = \rho\mathbf{v}$, $\dot{S}$ - the shear stress tensor. $P_t(t)$ is the space averaged pressure

$$P_t(t) = \frac{1}{\mathcal{V}}\int_{\mathcal{V}} p\mathcal{V}$$

so that the total pressure is

$$p = P_t(t) + p_d,$$

where p_d is the dynamic pressure and $grad\, p = grad\, p_d$. Thus the modified pressure Π is given by

$$\Pi = p_d + \frac{2}{3}\mu\, div\, \mathbf{v}.$$

Note that at known $\partial\rho/\partial t$ the system of equations (6)- (7) have the same behaviour in general as incompressible Navier-Stokes equations relative to the variables $\mathbf{V}$ and Π.

Boundary conditions are

$$\frac{\partial u}{\partial y} = \frac{\partial T}{\partial y} = \frac{\partial Y}{\partial y} = 0, \; v = 0 \text{ at the symmetric centerline,}$$

$$-\lambda\frac{\partial T}{\partial n} = \alpha(T - T_\infty), \quad \frac{\partial Y}{\partial n} = 0, \; u = v = 0, \; \text{at the solid walls,}$$

where T_∞ is the outside temperature, u and v are the axial and transversal components of velocity, n is the normal direction to the wall, α is the heat transfer coefficient of the wall.

The fictitious transport coefficients are:

$$\lambda^\epsilon(\mathbf{x}) = \begin{cases} \lambda(\mathbf{x}), \text{ if } \mathbf{x} \in \Omega, \\ \epsilon, \quad\quad \text{if } \mathbf{x} \in \Omega_f \end{cases},$$

$$D^\epsilon(\mathbf{x}) = \begin{cases} D(\mathbf{x}), \text{ if } \mathbf{x} \in \Omega, \\ \epsilon, \quad\quad \text{if } \mathbf{x} \in \Omega_f \end{cases},$$

$$\xi^\epsilon(\mathbf{x}) = \begin{cases} 1, \quad \text{ if } \mathbf{x} \in \Omega, \\ \epsilon^{-1}, \text{ if } \mathbf{x} \in \Omega_f \end{cases},$$

where Ω is the physical flow domain, Ω_f is the fictitious domain (for example, obstacle domain), ϵ is the small positive parameter, in the present calculation $\epsilon = 10^{-10}$.

At initial time the gas mixture is at rest at a temperature of $T_0 = 300\,K^0$ and pressure of $p = P_t(0) = P_0 = 10^5\,Pa, \; (p_d = 0)$.

A line igniter with circular cross section, placed at the centerline near the left side wall and perpendicular to the $x - y$ plane, initiates the combustion of the mixture:

$$S_{ig} = \begin{cases} Q_{ig}(1 - ((t - t_{ig})/t_{ig})^2), & \text{if } |\mathbf{x} - \mathbf{x_0}|^2 \leq r_{ig}^2 \text{ and } t \leq 2t_{ig} \\ 0, & \text{if } |\mathbf{x} - \mathbf{x_0}|^2 > r_{ig}^2 \text{ or } t > 2t_{ig} \end{cases}.$$

Here Q_{ig} is the igniter intensity, $\mathbf{x_0}$ is the coordinate of igniting zone center, r_{ig} is the radius of circular igniting zone, t_{ig} is the ignition period.

3 The numerical technique and problem parameters

3.1 The numerical technique

In the first step of the computation the transport equations (1)-(2) for enthalpy and mass fraction of methane are solved by a Crank-Nicolson scheme, then temperature is calculated from the equation (3). The average thermodynamic pressure is calculated from the ODE (4), obtained from integrating the mass conservation equation over the whole volume of the vessel. The density is obtained from the equation of state (5) and its value is corrected using the value of total gas mass in the vessel. Then the procedure is repeated to get a solution of non-linear equations (1)-(5).

In the closing step, velocities and the dynamic pressure field are found from equations (6) and (7) using a projection method for a known value of the density:

$$\frac{\mathbf{V}^{n+1} - \mathbf{V}^n}{\Delta t} = -grad\,\Pi^{n+1} + div(2\mu\dot{S}^n - \mathbf{v^n} \otimes \mathbf{V^n}) - \xi^\epsilon \mathbf{V^{n+1}},$$

$$\mathbf{V}^* = \mathbf{V}^n + \Delta t\, div(2\mu\dot{S}^n - \mathbf{v^n} \otimes \mathbf{V^n}),$$

$$\mathbf{V}^{n+1} = \psi^\epsilon(\mathbf{V}^* - grad\,\Pi^{n+1}),$$

where
$$\psi^\epsilon = (1 + \Delta t\,\xi^\epsilon(\mathbf{x}))^{-1} = \begin{cases} 1, & \text{if } \mathbf{x} \in \Omega \\ \frac{\epsilon}{1+\epsilon}, & \text{if } \mathbf{x} \in \Omega_f \end{cases}.$$

The equation for modified pressure field

$$\frac{\partial}{\partial x}\left(\eta^\epsilon \frac{\partial \Pi^{n+1}}{\partial x}\right) + \frac{\partial}{\partial y}\left(\eta^\epsilon \frac{\partial \Pi^{n+1}}{\partial y}\right) = f^\epsilon \qquad (8)$$

obtained using the mass conservation equation

$$div\,\mathbf{V}^{n+1} = -\frac{\rho^{n+1} - \rho^n}{\Delta t}.$$

Here
$$\eta^\epsilon(\mathbf{x}) = \begin{cases} 1, & \text{if } \mathbf{x} \in \Omega, \\ \epsilon, & \text{if } \mathbf{x} \in \Omega_f \end{cases},$$

$$f^\epsilon(\mathbf{x}) = \begin{cases} f^{n+1}, & \text{if } \mathbf{x} \in \Omega, \\ 0, & \text{if } \mathbf{x} \in \Omega_f \end{cases}.$$

The numerically boundary condition for Π is

$$\left.\frac{\partial \Pi}{\partial n}\right|_S = 0.$$

The equation for modified pressure (8) is solved by SOR-method.
The global accuracy of the solution is $O(\Delta t,\ \Delta x^2)$.

3.2 The problem parameters

The simulations are performed for the cases when five different types of obstacles mounted perpendicular to the axis of the vessel:
- a flat plate with a height of $h/2$ placed at the walls of the vessel at various distances from the left side wall (blockage ratio (BR) of obstacles is 0.5);
- a flat plate with a height of h placed at the center of the vessel at various distances from the left side wall (BR=0.5);
- a flat wire grid $40mm$ in height with a mesh width of $6mm$ and wire diameter of $2mm$ in the vessel with a length of $152mm$, and width of $40mm$ (BR=0.3);

- three flat plates with height h mounted at various distances x from the left side wall in the center of combustion chamber (BR=0.5);
- a flat plate with a height of $4mm$ placed at the walls and near the right side wall in the vessel with a length of $25mm$ and width of $5mm$ (BR=0.8).

In the calculation the vessel sizes, thermodynamic data, transport and kinetic coefficients are the following:

the sizes of vessel and obstacles are $L/l = 76$, $h/l = 10$, $l = 5 \times 10^{-4}\,m$;

the dynamic and thermodynamic data are $T_\infty = T_0 = 300\,K^0$, $\kappa_f = 1.25$, $\rho_0 = 1.107\,kg/m^3$, $W = 0.0276\,kg/mole$, $Pr = 0.73$, $c_{p0} = c_p(T_0) = 1128\,J/kg/K$, $e_{O_2} = 0.917$, $e_{CO_2} = 1.066$, $e_{H_2O} = 2.08$, $e_{CH_4} = 3$; $Nu_0 = l\alpha/\lambda_0 = 1$, $Re_0 = u_n l\rho_0/\mu_0 = 12.3$;

the transport coefficients are $\mu_0 = 1.8 \times 10^{-5}\,kg/m/s$, $\lambda_0 = \mu_0 c_{p0}/Pr$, $Le = \lambda/\rho D c_p = 1$;

the kinetic data are $u_n = 0.4\,m/s$, $l = 5\times 10^{-4}\,m$, $A = 2.4 \cdot 10^{10}\,m^3/mole \cdot sec$, $T_a = E/R = 24370\,K^0$, $Q = 5. \times 10^7\,J/kg$;

the igniter parameters are $Q_{ig} = 6.68 c_{p0} P_0 T_0 u_n/l$, $t_{ig} = 0.45 l/u_n$, $r_{ig} = 1.5l$, $x_{ig} = 0.7l$, $y_{ig} = 0$.

Here ρ_0 is the density of the initial unburned gas, Nu_0 is the Nusselt number, Re_0 is the Reynolds number.

The calculations are performed on a uniform mesh 200x1520 which provides a fine resolution of the flame front including preheat and reaction zones.

4 The results

Similar to the experiments of Starke [10] we will first consider the results obtained in cases when a flat plate placed at the walls of the vessel at four different distances x from the left end.

Figures 1, 2 and 3 demonstrate the influence of the obstacles on the flame propagation. Figure 1 displays the evolution of the reaction rate contours, Fig. 2 - the average pressure history in the vessel. Figure 3 display the reaction rate, velocity and vorticity fields near the refracting flame front.

Similar to experiments:

a real tulip formation occurs only when the obstacle is placed close to end flanges (Fig. 1, a and d);

a flame front refraction takes place in case b (flame shape between 6 and 8 positions, Fig. 1, b) and c (flame shape between 7 and 9 positions, Fig. 1, c);

a fastest mixture burning out occurs in case c then b (Fig. 2);

a strongest pressure jump takes place in case c (Fig. 1).

In addition the arising an isolated burning zone around the obstacle is established (Fig.1, b and c) and the baroclinic vorticity generating and increasing at the refracting flame is observed (Fig. 3).

Note that the sizes of the vessel are larger by ten times in the experiment than in the simulation. So, the combustion has become turbulent for cases b, c

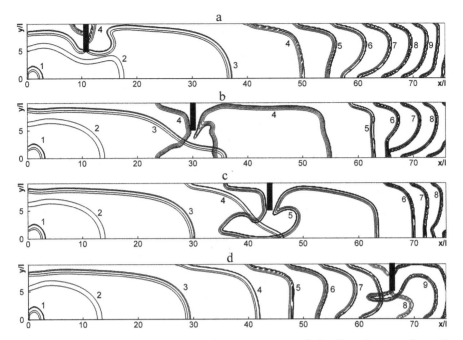

Fig. 1. Flame evolution for four different positions of the flat plate at the wall: $1 - 1ms$; $2 - 5ms$; $3 - 9ms$; $4 - 13ms$; $5 - 17ms$; $6 - 21ms$; $7 - 25ms$; $8 - 29ms$; $9 - 33ms$; $10 - 37ms$. Position of the obstacle: $a - x_{obs}/l = 11$; $b - x_{obs}/l = 30.4$; $c - x_{obs}/l = 44$; $d - x_{obs}/l = 66, l = 0.5mm$

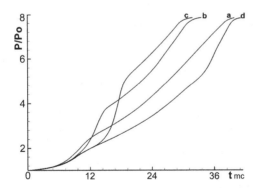

Fig. 2. Average pressure records in the vessel with the flat plate as an obstacle. Position of the obstacle: $a - x_{obs}/l = 11$; $b - x_{obs}/l = 30.4$; $c - x_{obs}/l = 44$; $d - x_{obs}/l = 66, l = 0.5mm$

in the experiment and the pressure jump is substantial higher in these cases than in simulation.

The following Fig. 4 demonstrate the influence of the flat plate placed at the center of the vessel on the propagation of the enclosed flame. As seen,

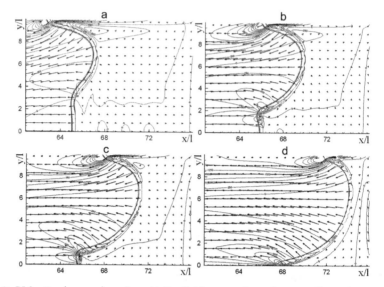

Fig. 3. Velocity (arrows) and vorticity fields near the refracting flame front in case of b (flame positions between $6-8$ in Fig.1). $t = ms$, $t = ms$, $t = ms$, $t = ms$

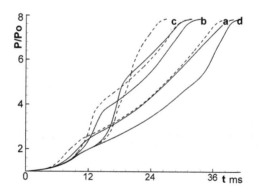

Fig. 4. Comparison of average pressure records of the flat plate at the wall (solid lines) and centered flat plate (dotted lines). Position of the obstacle: $a - x_{obs}/l = 11$; $b - x_{obs}/l = 30.4$; $c - x_{obs}/l = 44$; $d - x_{obs}/l = 66, l = 0.5mm$

there is no big differences in the average pressure records for these two cases except case c.

The propagation of flames through a circular wire grid mounted at $50mm$ from the point of ignition is illustrated in Fig.5. After the passing through grid the flame was split into three flamelets as number of circular wires. The burning rate increasing noticeably at this time (Fig. 8, c).

The Fig. 6 demonstrate the influence of the three flat plates mounted at various distances from the bottom side wall (from the ignition point) on the propagation of the enclosed flame. Similar to experiments of [11] the flame

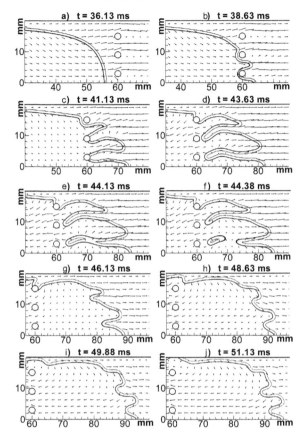

Fig. 5. Flame propagation through the circular wire grid mounted at $60mm$ from the point of ignition

flattening as it approaches the first obstacle, jetting past the obstacles, turning behind the obstacles, and the flame reconnection. Despite the fact that in the experiment flow was turbulent the flame shape around and behind the obstacles is predicted very well. In particular the three flat flames approaching each other behind the obstacles as in the experiment is well predicted.

Fig. 7 shows a flame propagation through the orifice when a flat plate with a height of $4mm$ placed near the right side wall. Blockage ratio of obstacle is 0.9. One can see the generation a flame jet coming to burned gas.

5 Conclusion

The fictitious domain method to detail simulation of flame propagation in the non-regular domain at low Mach number is elaborated. The computer code to predict of flame propagation in the vessel with non-regular configuration is

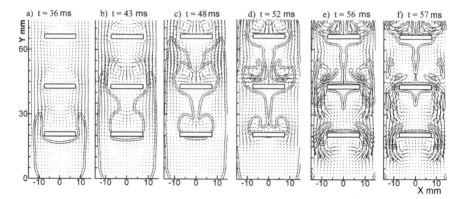

Fig. 6. Flame propagation through the three obstacles. Three flat plates with width $15mm$ mounted at distances $20mm$, $42.5mm$, $66mm$ from the point of ignition

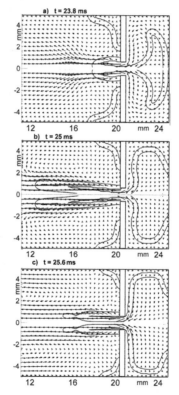

Fig. 7. Jet flame propagation through the orifice. A flat plate mounted at $21mm$ from the point of ignition

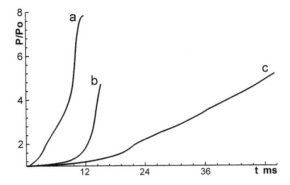

Fig. 8. Average pressure records for the cases: a - a flat plate placed near the right side wall; b - three flat plates mounted at various distances from the bottom side wall; c - a circular wire grid mounted at $50mm$ from the point of ignition

carried out and the calculation of flame propagation in the rectangular tube at various type of obstacles is implemented. Evolution of the average pressure in combustion chamber, velocity, vorticity, temperature, and dynamic pressure fields are found. The obtained results are compared with experimental data available in the literature and showing very well agreement with them. So, the efficiency of the fictitious domain method to simulate of reactive flow problem in non-regular domain is demonstrated.

6 Acknowledgments

This work supported by Kazakh Education and Science Ministry, Grant N 1090-2004 and partially sponsored by German Academic Exchange Service.

References

1. Smagulov Sh, Orunkhanov M (1981) Doklady AN USSR 260 (5):1078–1082 (in Russian)
2. Glowinski R, Pan W, Periaux J (1994) Comp Meth Appl Mech Eng 112:133–148
3. Collino F, Milot F (1997) J Comp Physics 138:907–938
4. Astrakhantsev A (1978) Zh Vychisl Mat Mat Fiz 18:118–125 (in Russian)
5. Glowinski R, Pan W, Periaux J (1995) Japan J Ind Appl Math 12:87–108
6. Glowinski R, Kearsley J, Pan W, Periaux J (1995) Int J Num Methods Fluids 20:695–711
7. Westbrook K, Dryer L (1981) Combust Sci Tech 27:31
8. Majda A, Sethian J (1985) Combust Sci Tech 42:185
9. Kaltayev A, Riedel U, Warnatz J (2001) Combust Sci Tech 158:53–69
10. Starke R, Roth P (1989) Combust Flame 75:111–121
11. Patel S, Ibrahim S, Yehia M, Hargrave G (2003) Exp Thermal Fluid Sci 27:355–361

Detailed numerical simulation of the auto-ignition of liquid fuel droplets

R. Stauch, S. Lipp, and U. Maas

Institute of Technical Thermodynamics, University of Karlsruhe(TH), Kaiserstraße 12, 76131 Karlsruhe, Germany stauch@itt.mach.uni-karlsruhe.de, lipp@itt.mach.uni-karlsruhe.de, maas@itt.mach.uni-karlsruhe.de

Summary. One and two dimensional numerical simulations of the auto-ignition process of single droplets of methanol and n-heptane in air are presented. Detailed models are used to simulate the transport processes as well as the chemical kinetics. Efficient numerical methods are implemented to reduce the computing time.

The influence of different ambient parameters on the ignition process is investigated. The ambient gas temperature turns out to be the physical parameter with the largest influence on the ignition delay time. With increasing ambient temperature the ignition delay time decreases. Furthermore, the ignition delay time decreases with increasing pressure following a power law. Two dimensional simulations show the almost exponential dependence of the ignition delay time on the velocity of a gas counterflow. If the counterflow is too strong, the flame is extinguished. Furthermore, the location of ignition is strongly affected by the counterflow velocity.

1 Introduction

A detailed understanding of droplet ignition and combustion is of interest in view of a reliable description of spray combustion. Especially a detailed understanding of the basic physical and chemical processes, like vaporization, transport and chemical kinetics, is required for reliable modelling. The simplest model of the fuel spray ignition process is the ignition of an ensemble of single fuel droplets. If microgravity droplet combustion is regarded, i.e. no gravitation, no relative motion of droplet and gas phase, the considered system can be assumed to have a spherical symmetry. Hence only a system of one-dimensional conservation equations has to be solved. This regime is appropriate to investigate the basic physical and chemical processes, like vaporization, molecular transport and chemical kinetics and their interaction. Particularly for describing transient processes like the droplet ignition the understanding of this interaction is necessary. To account for a gas flow around the droplet or the ignition process of an array of droplets at least two-dimensional geometries have to be simulated.

The combustion of a single liquid droplet in a quiescent atmosphere has been studied numerically. Numerical simulations have been presented for methanol droplets [1, 2, 3] and n-heptane droplets [4, 5, 6, 7]. Only a little number of studies do not assume spherical symmetry and nevertheless consider the physics of the droplets and the chemical processes in detail [8, 9, 10, 11, 12, 13]. Few studies focus on the ignition process of fuel droplets of higher hydrocarbons [5, 6, 7, 14, 15, 16, 17, 18]. Takei et al. [14], Nakanishi et al. [16] and Tanabe et al. [17] have determined ignition delay times of droplets experimentally. Tsukamoto et al. account for chemical kinetics by a one-step irreversible overall reaction [15]. Schnaubelt et al. investigate the ignition process of n-heptane and n-decane experimentally and numerically [6, 18]. Moriue et al. simulate the ignition of a fuel droplet in a closed volume [7]. However, the influence of the ambient gas temperature and the ambient pressure on the ignition process in the case of auto-ignition has not been investigated extensively based on detailed numerical simulations. With respect to technical applications, like gas turbines or combustion engines, the influence of ambient physical properties on the ignition process are of major interest. Parametric studies are performed to construct libraries of droplet combustion which can be used, e.g. in flamelet-like calculations of turbulent spray combustion. Parameters like the ambient gas temperature, pressure as well as the velocity of a gas counterflow are varied. In the case of the ignition of a droplet pair simulations with different droplet distances are performed. Studies of the ignition and combustion of methanol and n-heptane droplets in air are presented.

2 Mathematical Model

The presented model describes a fuel droplet surrounded by an ambient gas phase. This allows to formulate the governing set of compressible Navier-Stokes equations for a reactive system and equations of state for the gas phase and the liquid phase [19].

$$\frac{\partial \rho}{\partial t} + \text{div}(\rho \mathbf{v}) = 0, \tag{1}$$

$$\frac{\partial \rho_i}{\partial t} + \text{div}(\rho_i \mathbf{v}) + \text{div}\mathbf{j}_i = M_i \dot{\omega}_i, \tag{2}$$

$$\frac{\partial (\rho \mathbf{v})}{\partial t} + \text{div}(\rho \mathbf{v} \otimes \mathbf{v}) + \text{div}\bar{\bar{p}} = 0, \tag{3}$$

$$\frac{\partial \rho u}{\partial t} + \text{div}(\rho u \mathbf{v} + \mathbf{j}_q) + \bar{\bar{p}} : \text{grad}\mathbf{v} = 0, \tag{4}$$

$$\text{gas}: \quad p = \frac{\rho}{M}RT, \qquad \text{liquid}: \quad \rho = \rho(T). \tag{5}$$

t denotes the time, ρ the density, $\mathbf{v}$ the velocity, ρ_i the density of species i, $\mathbf{j}_i$ the diffusion flux density of species i, M_i the molar mass of species i, $\dot{\omega}_i$ the molar scale rate of formation of species i, $\bar{\bar{p}}$ the pressure tensor, u the specific inner energy, $\mathbf{j}_q$ the conductive heat flux density, $\bar{M}$ the mean molar mass and R the universal gas constant.

The chemical kinetics of methanol is modelled by a detailed reaction mechanism of Chevalier and Warnatz [20], containing 23 chemical species and 166 elementary reactions. For n-heptane a detailed reaction mechanism of Golovitchev [21], containing 62 chemical species and 572 elementary reactions is used. The transport processes are also modelled in detail. Fourier's law is used to determine the heat fluxes. For the determination of the diffusion coefficients the approximation of Curtis and Hirschfelder [22] is used. Convection inside the droplet is neglected. The liquid phase properties are calculated based on the data correlations taken from Reid et al. [23]. The approximation of Latini et al. is used to calculate the heat conductivities, the approximation of Rowlinson and Bondi to calculate the specific heat capacities. The necessary properties to model the phase transition are also taken from Reid et al. [23]. The vapor pressure is calculated using the Wagner equation, and the enthalpy of vaporization is calculated by the correlation of Riedel and Watson.

A vaporization model accounts for the coupling of the liquid phase and the gas phase. A local phase equilibrium is modelled by interface equations.

$$\phi_{\text{vap}} = \frac{-\sum\limits_{j_{vap}} j_j^g - \sum\limits_{j_{vap}} w_j \sum\limits_i R_i + \sum\limits_{j_{vap}} R_j}{\sum\limits_{j_{vap}} \frac{p_j M_j}{pM} - 1}, \tag{6}$$

$$0 = \rho \cdot v_n - \phi_{\text{vap}} - \sum_i R_i, \tag{7}$$

$$0 = \phi_{\text{vap}} (w_i - \epsilon_i) + j_i + w_i \sum_j R_j - R_i, \tag{8}$$

$$0 = \sum_i \epsilon_i \cdot \phi_{\text{vap}} \cdot \Delta h_{\text{vap},i} + j_q^g - j_q^l. \tag{9}$$

ϕ_{vap} denotes the vaporization rate, w_i the mass fraction of species i, R_i the surface reaction rate of species i, j_{vap} the index of the vaporizing species, p_j the partial pressure of species j, p the pressure, M_j the molar mass of species j, $\bar{M}$ the mean molar mass, ρ the density, v_n the normal velocity, $\epsilon_i = \dot{m}_i/\dot{m}$ the fraction of vaporizing mass, j_i the diffusion flux density of species i (gas phase), $\Delta h_{vap,i}$ the enthalpy of vaporization of species i and j_q the heat flux density.

In the spherically symmetric case (1D) the equation system is transformed into modified Lagrangian coordinates to overcome difficulties with the discretization of the convective terms [24]. This allows to implement a low Mach number approximation easily and to replace the momentum equation by $\frac{\partial p(r)}{\partial r} = 0$. The resulting equations are discretized by central differences on a non-equidistant grid.

Fig. 1. Transformation into boundary-fitted coordinates

In the case of droplets in a convective environment (2D) the governing equations are transformed into generalized coordinates $\xi(x, y)$, $\eta(x, y)$. This boundary-fitted curvilinear coordinate system is generated by the grid generator TOMCAT by solving a Poisson equation system after specification of the boundaries of the grid [25].

The spatial discretization is done by the method of lines using finite difference techniques. For the inviscid fluxes the flux vector splitting upwind scheme of Steger and Warming [26] is used. The flux vectors are split into subvectors by similarity transformations. The viscous fluxes are discretized using second order central differences in chain rule conservation law form [27].

In both cases (1D and 2D) one obtains a large system of ordinary differential and algebraic equations. The resulting differential-algebraic equation system is solved by the linearly implicit extrapolation method LIMEX [28]. This method was conceived for the solution of such large and stiff differential-algebraic systems. In addition, efficient numerical methods are implemented for the calculation of the block tridiagonal and block nonadiagonal matrices and the solution of the linear equation systems [29]. Due to simultaneous perturbation of the sparse Jacobian the needed number of function evaluations does not depend on the number of grid points.

The program code is written in FORTRAN. Computations are performed on a regular single PC using LINUX as operating system. The simulations yielding to the presented results use the full capacity of the mentioned hardware. One dimensional simulations last for approximately ten CPU hours on grid with 50 points. Two dimensional simulations are performed on a grid with 40x20 points lasting forty CPU hours. Simulations of more complex systems are possible, because the program code can be easily parallelized by an appropriate compiler.

Results and Discussion

The knowledge of ignition delay times in the case of auto-ignition is of importance for many practical applications, e.g. Diesel engines or Lean Premixed Prevaporized (LPP) gas turbines [30]. First, the ignition delay times of single fuel droplets in a stagnant gas surrounding are presented for the fuels methanol and n-heptane. Subsequently studies of the ignition process of methanol droplets in a gas counterflow are presented.

2.1 Ignition of fuel droplets in a stagnant gas phase

Figure 2 shows the dependence of the ignition delay time of single methanol droplets on the ambient gas phase temperature, i.e. the parameter with the largest influence on the ignition delay time. The ignition delay times are presented for the temperature range from 1000K to 2000K.

Figure 2 shows the typical exponential dependence of the ignition delay time on the ambient temperature for a homogeneous stoichiometric methanol/air mixture. In the case of the ignition of a single methanol droplet the ignition delay time also decreases exponentially with increasing temperature. Though, the decrease is not that strong as in the case of the homogeneous gas phase. The flattening of the temperature dependence is originated from the physical transport processes. With increasing temperature the physical transport speeds up not as rapidly as the chemical kinetics.

Because it is hard to control and to measure the initial conditions of single droplet experiments, simulations are performed for two different initial conditions in order to yield information on the sensitivity of the ignition delay time relating to the initial conditions. On the one hand no initial gaseous n-heptane is present (INI1), on the other hand simulations start with a small amount of prevaporized n-heptane in the gas phase (INI2).

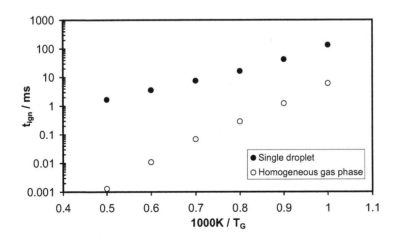

Fig. 2. Arrhenius plot of the dependence of the ignition delay time on ambient gas temperature (methanol, $p = 7$bar, single droplets $r_D = 200\mu$m, homogeneous gas phase $\lambda = 1$)

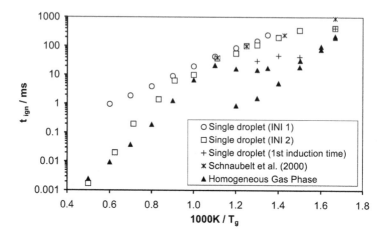

Fig. 3. Arrhenius plot of the dependence of the ignition delay time on ambient gas temperature (n-heptane, $p = 7$bar, single droplets $r_D = 200\mu$m, homogeneous gas phase $\lambda = 1$)

Figure 3 shows the dependence of the ignition delay time on the ambient gas phase temperature in the case of n-heptane droplets for a temperature range from 600K to 2000K with an ambient pressure of 7bar and a droplet radius of 200μm. Simulations are performed with the two different initial conditions (INI1, INI2) mentioned above. Between 650K and 850K a two-stage ignition of the n-heptane droplet can be observed. Hence, not only the total ignition delay time but the first induction time are shown in figure 3. Additionally, the first and total ignition times of a homogeneous stoichiometric gaseous n-heptane/air mixture are included.

The typical temperature dependence of the ignition delay times can be seen for the case of a homogeneous n-heptane/air mixture. The ignition delay time depends strongly on the temperature below 700K and above 900K. Between these two temperatures the ignition delay time increases with increasing temperature (NTC-behaviour). This behaviour of the total ignition delay times cannot be observed in the case of the igniting droplet. Solely the first induction time of the droplets shows nearly no temperature dependence (ZTC-behaviour) in the range of 650K to 760K. For temperatures below 1000K the temperature dependence appears to be almost the same for both initial conditions. Above 1000K the dependence of the ignition delay time on the ambient temperature is stronger in the case of prevaporized n-heptane (INI2) than in the case of no initial gaseous n-heptane (INI1). This fact leads to much shorter ignition delay times for higher temperatures in the case of INI2. So the physical transport processes are determined to be rate limiting in this temperature range. In addition, ignition delay times of n-heptane droplets

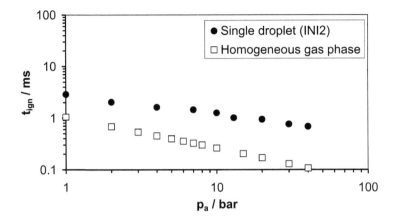

Fig. 4. Dependence of ignition delay time on ambient pressure (n-heptane, $T_g = 1200K$, single droplets $r_D = 50\mu m$, homogeneous gas phase $\lambda = 1$)

obtained by Schnaubelt et al. [18] are shown in figure 3. These ignition delay times show the same qualitative behaviour, whereas it should be noted that these results refer to different ambient conditions ($r_D = 350\mu m$, $p = 5bar$).

Not only the dependence of the ignition delay on the ambient temperature, but also the dependence of the ignition delay time on the ambient pressure is of great interest. In figure 4 the pressure dependence of the ignition delay time is shown for the case of an n-heptane droplet in air with a droplet radius of $r_D = 50\mu m$ and a gas temperature of $T_g = 1200K$. Additionally, again the pressure dependence of a homogeneous stoichiometric gaseous n-heptane/air gas mixture is included.

A decrease of the ignition delay time of an n-heptane droplet with increasing pressure can be observed. This decrease follows a power law with the exponent -0.4 (note, that the scaling of both axes of this figure is logarithmic).

$$t_{\mathrm{ign}}(p) \propto \left(\frac{p}{p_0}\right)^{-0.4}. \tag{10}$$

Other authors [15, 16] have determined almost the same behaviour. In the case of the homogeneous n-heptane/air mixture the qualitatively same behaviour is obtained. However, the decrease of the ignition delay time with increasing pressure is stronger, which results in another exponent of the power law of approximately -0.6.

2.2 Ignition of fuel droplets in a convective gas phase

In the following the influence of a gas counterflow on the ignition process of a methanol droplet is investigated. To study the sensitivity of this dependence on the initial conditions, again, simulations are performed with the two different initial conditions of no initial gaseous methanol (INI1) and prevaporized methanol in the gas phase (INI2).

The dependence of the ignition delay time on the velocity of a stationary gas counterflow is presented in figure 5. No initial gaseous methanol is present (INI1), the droplet radius is 200μm and the gas temperature $T_g = 1500K$.

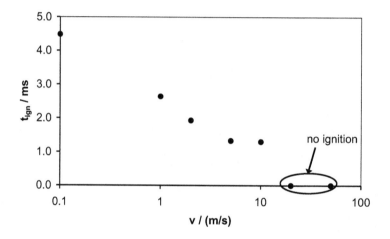

Fig. 5. Dependence of ignition delay time on droplet velocity (methanol, INI1, $r_D = 200\mu m$, $T_g = 1500K$)

The ignition delay time decreases with increasing counterflow velocity. Compared to the case of a stagnant gas environment the ignition delay time for a counterflow velocity of 10m/s is reduced by a factor of 4. With an increasing droplet velocity the profiles are steepened and accordingly the diffusion fluxes become higher. The higher diffusion fluxes accelerate the processes of mixing and heat conduction. Therefore, a faster formation of an ignitable mixture at a sufficiently high temperature is promoted and the ignition delay time decreases. For velocities above 20m/s no ignition occurs any more. There, the local strain is so high, that the flame extinguishes.

In figure 6, one can see the dependence of the ignition delay time on the counterflow velocity, if prevaporized methanol in the gas phase exists.

For counterflow velocities below 20m/s the ignition delay time remains almost constant despite the change of the velocity by a factor of 1000. This behaviour is contrary to the case of no initial gaseous methanol (INI1), where

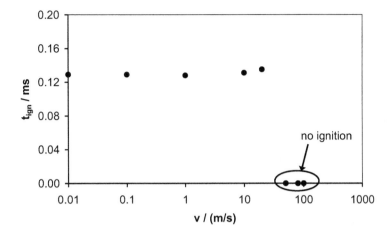

Fig. 6. Dependence of ignition delay time on droplet velocity (methanol, INI2, $r_D = 200\mu$m, $T_g = 1600$K)

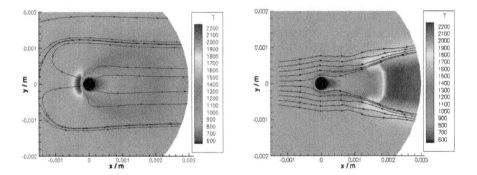

Fig. 7. Temperature profiles during the ignition process (methanol, $p = 7$bar, $r_D = 200\mu$m, $T_g = 1500$K): left: upstream ignition ($v = 1$m/s); right: downstream ignition ($v = 10$m/s)

the ignition delay time is strongly affected by the gas counterflow velocity. The reason of this behaviour is supposed to be the fact, that if prevaporized methanol exists the transport processes are not that time-limiting like in the case of no initial gaseous methanol. If the velocity is larger than 50m/s no ignition occurs.

Other characteristical properties of the ignition and combustion process are also affected by the gas counterflow. In figure 7 the temperature profiles are shown for a lower velocity of 1m/s and a higher velocity of 10m/s. at an early stage of ignition. In the case of the lower velocity the ignition process is initiated upstream, very close to the droplet surface. In contrast, the ignition occurs in the wake of the droplet for higher counterflow velocities.

3 Conclusions

The auto-ignition process of isolated fuel droplets in air is investigated. The simulations are performed for one and two dimensional geometries and with detailed models for the vaporization, transport process and chemical kinetics. The numerical model of the simulations is given in detail. Efficient numerical methods are devised to minimize computing time.

The influence of different ambient parameters on the ignition delay time is studied. The ignition delay time shows a strong dependence on the ambient gas temperature. The pressure dependence of the ignition delay time follows a power law with an exponent of -0.4. Two dimensional simulations show the influence of a gas counterflow on the ignition process. With increasing counterflow velocity the ignition delay time decreases. In addition, the location of ignition is strongly affected by the gas counterflow. Though, the influence of almost all ambient parameters also depends on the initial conditions.

Further studies will focus on the auto-ignition process of multicomponent droplets, like n-heptane/iso-octane droplets, to achieve a better description of technical conditions. Two dimensional simulations will deal with the ignition process of droplet pairs.

4 Acknowledgments

The authors wish to thank the DFG for financial support in the frame of SFB 606.

References

1. Cho SY, Choi MY, Dryer FL (1990) Proc Combust Inst 23:1611–1617
2. Cho SY, Yetter RA, Dryer FL (1992) J Comput Physics 102:160–179
3. Stapf P (1992) Modellierung der Tröpfchenverbrennung unter Einschluß detail-lierter chemischer Reaktion. PhD thesis. Universität Stuttgart
4. Cho SY, Dryer FL (1999) Combust Theory Model 3:267–280
5. Marchese AJ, Dryer FL, Nayagam V (1999) Combust Flame 116:432–459
6. Schnaubelt S, Moriue O, Eigenbrod C, Rath HJ (2001) Microgravity Sci Technol XIII/1:20–23
7. Moriue O, Mikami M, Kojima K, Eigenbrod C (2005) Proc Combust Inst 30: 1973–1980
8. Tsai JS, Sterling AM (1990) Proc Combust Inst 23:1405–1412

9. Dwyer HA, Aharon I, Shaw BD, Niazmand H (1996) Proc Combust Inst 26:1613–1619
10. Stapf P, Dwyer HA, Maly RR (1998) Proc Combust Inst 27:1857–1864
11. Okai K, Moriue O, Araki M, et al (2000) Combust Flame 121:501–512
12. Aouina Y, Maas U, Gutheil E, Riedel U, Warnatz J (2001) Combust Sci Technol 173:1–23
13. Shaw BD, Dwyer HA, Wei JB (2002) Combust Sci Technol 174:29–50
14. Takei M, Tsukamoto T, Niioka T (1993) Combust Flame 93:149–156
15. Tsukamoto T, Okada H, Niioka T (1993) Trans Japan Soc Aeronaut Space Sci 35:165–176
16. Nakanishi R, Kobayashi H, Kato S, Niioka T (1994) Proc Combust Inst 25:447–453
17. Tanabe M, Kono M, Sato J, et al (1995) Combust Sci Technol 108:103–119
18. Schnaubelt S, Moriue O, Coordes T, Eigenbrod C, Rath HJ (2000) Proc Combust Inst 28:953–960
19. Hirschfelder JO, Curtiss CF (1949) Proc Combust Inst 3:121–127
20. Chevalier C (1993) Entwicklung eines detaillierten Reaktionsmechanismus zur Modellierung der Verbrennungsprozesse von Kohlenwasserstoffen bei Hoch- und Niedertemperaturbedingungen. PhD thesis. Institut für Technische Verbrennung, Universität Stuttgart
21. Golovitchev V (2004) http://www.tfd.chalmers.se /~valeri/MECH.html
22. Hirschfelder JO, Curtiss CF, Bird RB (1964) Molecular theory of gases and liquids. John Wiley & Sons, New York:
23. Reid RC, Prausnitz JM, Poling BE (1989) The properties of gases and liquids. McGraw-Hill
24. Stauch R, Lipp S, Maas U (2005) Detailed numerical simulations of the auto-ignition of single n-heptane droplets in air (submitted)
25. Thompson JF, Thames FC, Mastin CW (1977) J Comput Physics 24:274–302
26. Steger JL, Warming RF (1981) J Comput Physics 40:263–293
27. Riedel U, Maas U, Warnatz J (1993) Comput Fluids 22:285–294
28. Deuflhard P, Hairer E, Zugck J (1987) Numerische Mathematik 51:501–516
29. Riedel U, Maas U, Warnatz J (1993) Impact Comput Sci Eng 5:20–52
30. Golovitchev V, Nordin N, Jarnicki R, Chomiak J (2000) SAE Paper 2000-01-1891

Numerical investigation of a supersonic flow with jet injection

A.Zh. Naimanova

Institute of Mathematics, Pushkin str. 125, 480100 Almaty, Kazakhstan
ked@math.kz

Introduction

One of the most important applied problems is the problem on calculation of supersonic turbulent jets in a cocurrent flow with subsonic domains, for instance, a problem on the outflow of a supersonic jet into a supersonic flow with partially limited domain or into a subsonic cocurrent flow. The main difficulties of modelling such flows are the laminar–turbulent transition, interaction between shocks, interaction of Mach waves with boundary layer and also the presence of separation zones [1–3]. It is more difficult to describe the above–mentioned processes on the basis of common numerical algorithm without application of various simplifying assumptions [1].

In the numerical investigations of supersonic jets the given systems of equations has a hyperbolic–parabolic type in supersonic domains. In subsonic domains there is the element of ellipticity due to the longitudinal pressure gradient and there is the necessity for some regularizations in order to ensure the hyperbolicity–parabolicity of the system. These problems limit the possibility of an effective use of the methods of numerical analysis.

In the present paper we consider the outflow of the system of three–dimensional supersonic jets into a cocurrent supersonic flow with partially limited domain and into a cocurrent subsonic flow.

1 Formulation of problem

From the system of circular nozzles of the identical size the supersonic turbulent jets flow out with the velocity u_0 into a concurrent supersonic flow with a partially limited domain, i.e. with walls at $y = 0$ and $y = 2L$ (fig. 1a), or into a free cocurrent subsonic flow, which moves with the velocity u_∞ (fig. 1b). The nozzles are located with some constant intervals. Due to the symmetry of nozzle placement with respect to y and z axis, the processes of interaction of flows in the shaded rectangle $OKMN$ (Ω) are examined.

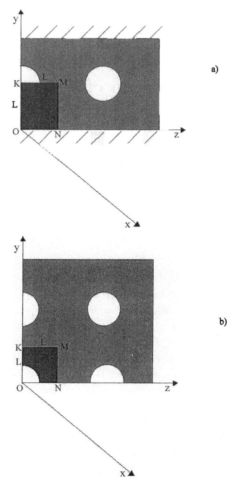

Fig. 1. Flow diagram

The system of averaged parabolized Navier-Stokes equations can be written down in the conservative form in the Cartesian coordinates as follows:

$$\frac{\partial \mathbf{E}}{\partial x} + \frac{\partial (\mathbf{F} - \mathbf{F}_v)}{\partial y} + \frac{\partial (\mathbf{G} - \mathbf{G}_v)}{\partial z} = 0, \tag{1}$$

$$\mathbf{E} = \begin{pmatrix} \rho u \\ \rho u^2 + p \\ \rho uv \\ \rho uw \\ (E_t + p)\, u \end{pmatrix}, \quad \mathbf{F} = \begin{pmatrix} \rho v \\ \rho uv \\ \rho v^2 + p \\ \rho vw \\ (E_t + p)\, v \end{pmatrix}, \quad \mathbf{G} = \begin{pmatrix} \rho w \\ \rho uw \\ \rho vw \\ \rho w^2 + p \\ (E_t + p)\, w \end{pmatrix},$$

$$\mathbf{F}_v = \frac{1}{Re}\left(0, \ \mu\frac{\partial u}{\partial y}, \ \frac{4}{3}\mu\frac{\partial v}{\partial y}, \ \mu\frac{\partial w}{\partial y}, \ u\mu\frac{\partial u}{\partial y} + \frac{4}{3}v\mu\frac{\partial v}{\partial y} + w\mu\frac{\partial w}{\partial y} + \right.$$

$$\left. +\frac{k}{(\gamma-1)\,M_a^2\,\mathrm{Pr}}\frac{\partial T}{\partial y}\right)^T,$$

$$\mathbf{G}_v = \frac{1}{Re}\left(0, \ \mu\frac{\partial u}{\partial z}, \ \mu\frac{\partial v}{\partial z}, \ \frac{4}{3}\mu\frac{\partial w}{\partial z}, \ u\mu\frac{\partial u}{\partial z} + v\mu\frac{\partial v}{\partial z} + \frac{4}{3}w\mu\frac{\partial w}{\partial z} + \right.$$

$$\left. +\frac{k}{(\gamma-1)\,M_a^2\,\mathrm{Pr}}\frac{\partial T}{\partial z}\right)^T,$$

$$p = (\gamma-1)\left[E_t - \frac{1}{2}\left(\rho u^2 + \rho v^2 + \rho w^2\right)\right], \quad c_v = \frac{1}{\gamma\,(\gamma-1)\,M_a^2},$$

$$T = \left(\frac{1}{\rho\,c_v}\right)\left[E_t - \frac{1}{2}\left(\rho u^2 + \rho v^2 + \rho w^2\right)\right], \quad \mu = \mu_l + \mu_t.$$

The system (1) is written in a dimensionless form using conventional symbols. The parameters at the nozzle (u_0, ρ_0, T_0) are taken as the determining parameters and the radius r of the edge is treated as a characteristic dimension. The pressure and the total energy are represented by $\rho_0\,u_0^2$.

Here $\gamma = c_p/c_v$ is specific heat ratio , c_p , c_v are specific heats at constant pressure and volume respectively, μ_t is the turbulent viscosity coefficient, M_a is the jet Mach number, Pr is the Prandtl number, μ_l is molecular viscosity coefficient, which is calculated by the Sutherland formula: $\mu_l = T^{3/2}\left(\frac{1+S_1}{T+S_1}\right)$, where $S_1 = 110\ K/T_0$. The gas is assumed to be perfect with a specific heat ratio $\gamma = 1.4$.

System (1) is closed by the Baldwin-Lomax algebraic model of turbulence. According to [4] the turbulent viscosity coefficient is defined as follows:

near the wall: $\mu_t = \rho l^2|\zeta|$, where ζ is the vorticity , $l = \chi y\left[1 - e^{-y^+/A}\right]$ is the universal variable, u_τ is the dynamic velocity, $A = 26$;

far from the wall: $\mu_t = 0.0168\,\rho V_0 L_0$, where $V_0 = \min\left[F_{\max}; \frac{0.25 q_{dif}^2}{F_{\max}}\right]$,

$L_0 = 1.6 y_{\max} I^k$, $F_{\max} = \max\left(|\zeta|\frac{l}{\chi}\right)$, $q_{dif} = u_{\max} - u_{\min}$,

$I^k = \left[1 + 5.5\left(\frac{0.3z}{y_{\max}}\right)^6\right]^{-1}$ — the Klebanov limiting multiplier, $y_{\max}$ corresponds to $F_{\max}$.

2 Boundary conditions

We perform the calculations starting from the outer edge $y = \delta_B$ of the viscous sublayer, in which we set

$$\tau = \tau_w, \quad q = q_w, \quad v = w = 0, \quad \frac{\partial p}{\partial y} = 0, \quad 0 \le z \le L, \quad x > 0, \quad (2)$$

where τ and q are the turbulent friction stresses and heat flux, respectively, and τ_w and q_w are their values at the wall, $\delta_B = y^+/(u_\tau \mathrm{Re})$ is the distance from the wall to the turbulent flow core, y^+ is the universal variable, $u_\tau = \left(\frac{1}{2}C_f\right)^{1/2} u_\infty$ is dynamic velocity, C_f is flow friction coefficient, $u_\infty = \frac{M_\infty}{M_a}\sqrt{T_\infty}$ and ∞ are values of the cocurrent flow parameters.

The turbulent friction stress in (2) is determined by the Prandtl formula [5] $\tau = \rho l^2 \left(\frac{\partial u}{\partial y}\right)^2$, where $l = \chi y$ is the mixing length, $\chi = 0.41$ is the Karman constant, and the wall friction stress $\tau_w = \frac{1}{2}\rho u_\infty^2 C_f$. Then the first equation in (2) can be written as follows

$$\left[\frac{\partial u}{\partial y}\right]_{y=\delta_B} = \frac{\mathrm{Re}}{\chi y^+} u_\tau^2, \quad 0 \le z \le L_1, \quad x > 0,$$

which is the boundary condition for the longitudinal velocity component.

The expression for the turbulent heat flux is $q = -\lambda \left(\frac{\partial T}{\partial y}\right)$.

The Reynolds analogy [5] relating friction and heat transfer has the form:

$$q_w = \rho c_p u_\infty (T_w - T_\infty) \frac{C_f}{2},$$

here, the pipe drag coefficient $\lambda = \rho \nu_{tw} c_p$ ($\nu_{tw} = \chi y u_\tau$ is the kinematic turbulent viscosity coefficient on the wall). The boundary condition for the temperature can be written as follows:

$$\left[\frac{\partial T}{\partial y}\right]_{y=\delta_B} = \frac{\mathrm{Re}}{\chi y^+} \frac{u_\tau^2}{u_\infty} (T_w - T_\infty), \quad 0 \le z \le L, \quad x > 0.$$

On the axis of symmetry the boundary conditions have the form:

$$\frac{\partial u}{\partial z} = \frac{\partial v}{\partial z} = \frac{\partial \rho}{\partial z} = \frac{\partial T}{\partial z} = 0, \quad w = 0, \quad z = 0,$$

$$z = L, \quad 0 \le y \le L, \quad x > 0,$$

$$\frac{\partial u}{\partial y} = \frac{\partial w}{\partial y} = \frac{\partial \rho}{\partial y} = \frac{\partial T}{\partial y} = 0, \quad v = 0, \quad y = L, \quad 0 \le z \le L, \quad x > 0,$$

where L is the transverse dimension of the considered domain in the directions y and z.

The boundary conditions in the inlet cross – section (the initial conditions) are given as follows.

In the jet

$$u = 1, \quad v = w = 0, \quad T = 1, \quad \rho = 1,$$

In the cocurrent flow

$$T = 1, \quad u = u_\infty, \quad v = w = 0, \quad p = \frac{1}{\gamma M_a^2 n}.$$

In the cocurrent flow in the neighbourhood of the wall the velocity can be described by the power law [6]:

$$u = u_\infty \left(\frac{y}{\delta}\right)^{1/7}, \quad v = w = 0, \quad \delta_B \leq y \leq \delta, \quad 0 \leq z \leq L,$$

where $\delta = 0.0586 \, (u_\infty \mathrm{Re})^{-1/5}$ is the turbulent boundary layer thickness.

The temperature field is given by the temperature dependence on the velocity for the case $T = T_w$ [6] and the pressure is defined as $\frac{\partial p}{\partial y} = 0$.

3 Method of solution

The presence of the derivative of the pressure with respect to the longitudinal coordinate in the system (1) creates the possibility of upstream perturbation propagation along the flow through the thin boundary layer. The solution of this problem obtained by the marching method can be divergent. Several ways of eliminating these ever-growing solutions are suggested in [7, 8]. By analogy with [8], in the given paper the flux vector is presented as the sum of two vectors $\mathbf{E} = \mathbf{E}^* + \mathbf{E}^p$:

$$\mathbf{E}^* = \begin{pmatrix} \rho u \\ \rho u^2 + \omega p \\ \rho uv \\ \rho uw \\ (E_t + p) u \end{pmatrix}, \quad \mathbf{E}^p = \begin{pmatrix} 0 \\ (1 - \omega) p \\ 0 \\ 0 \\ 0 \end{pmatrix}, \tag{3}$$

The parameter ω is defined from the analysis of eigen-functions:

$$\omega = \begin{cases} 1, & \text{with} \quad M_x \geq 1 \\ \sigma \gamma M_x^2, & \text{with} \quad M_x < 1 \end{cases},$$

where $\sigma \approx 0.71$ is the safety factor, $M_x = u/a$ is the local Mach number and a is the speed of sound.

Taking into account (3) the system of the equations (1) can be written in the form:

$$\frac{\partial \mathbf{E}^*}{\partial x} + \frac{\partial \mathbf{E}^p}{\partial x} + \frac{\partial (\mathbf{F} - \mathbf{F}_v)}{\partial y} + \frac{\partial (\mathbf{G} - \mathbf{G}_v)}{\partial z} = 0. \tag{4}$$

The order of definition of unknown values can be represented in the form of a two–step splitting scheme (on the first step it is assumed, that the transfer of fluxes is carried out by the convection, in second step – by the diffusion): The first step – the calculation of intermediate values of fluxes:

$$\frac{\mathbf{E}^{*i} - \mathbf{E}^{*n}}{\Delta x} = -\frac{\partial \mathbf{F}^n}{\partial y} - \frac{\partial \mathbf{G}^n}{\partial z} - \frac{\partial \mathbf{E}^p}{\partial x}. \tag{5}$$

The second step – the calculation of final values of the functions:

$$\frac{A^n \left(\mathbf{U}^{n+1} - \mathbf{U}^{*i}\right)}{\Delta x} = \frac{\partial \mathbf{F}_v^{n+1}}{\partial y} + \frac{\partial \mathbf{G}_v^{n+1}}{\partial z}, \tag{6}$$

where $\mathbf{U} = [\rho,\ \rho u,\ \rho v,\ \rho w,\ E_t]^T$, $A = \frac{\partial \mathbf{E}^*}{\partial \mathbf{U}}$ is the Jacobian matrix shown in [8].

The numerical solution (5) with respect to the flux $\left(\mathbf{E}^{*i}\right)$ is done by the MacCormack explicit scheme [8] ($i = 2$), which has a second order of accuracy on spatial variables, or the Warming-Kutler-Lomax three-step scheme [8] ($i = 3$) with the third order of accuracy. After the calculation of the vector $\mathbf{E}^{*i}$, from the solution of five nonlinear algebraic systems of the equations, obtained from the first expression (3), the components of the vector $\mathbf{U}^{*i}$ are defined (for simplicity we omit the upper index). For the supersonic field these components can be written in the following form:

$$\begin{pmatrix} U_2 \\ \frac{U_2^2}{U_1} + \left(U_5 - \frac{U_2^2+U_3^2+U_4^2}{2U_1}\right)(\gamma-1) \\ \frac{U_2 U_3}{U_1} \\ \frac{U_2 U_4}{U_1} \\ \left(U_5 + (\gamma-1)\left(U_5 - \frac{U_2^2+U_3^2+U_4^2}{2U_1}\right)\right)\frac{U_2}{U_1} \end{pmatrix} = \mathbf{E}^{*i}; \quad \mathbf{E}^{*i} = \begin{pmatrix} E_1 \\ E_2 \\ E_3 \\ E_4 \\ E_5 \end{pmatrix}, \tag{7}$$

The solution of the system (7) has the form:

$$U_2 = E_1 \quad, \quad U_3 = \frac{E_3 U_1}{E_1} \quad, \quad U_4 = \frac{E_4 U_1}{E_1} \quad,$$

$$U_5 = \frac{1}{U_1}\left(\frac{E_5 U_1^2}{E_1} - \left(E_2 U_1 - E_1^2\right)\right),$$

$$U_1 = \frac{\gamma E_2 E_1^2 - E_1^2 \sqrt{(\gamma E_2)^2 - (\gamma^2 - 1)\left(2E_1 E_5 - E_3^2 - E_4^2\right)}}{(\gamma-1)\left(2E_5 E_1 - E_3^2 - E_4^2\right)}.$$

In (7) the discriminant is equal to zero for the case $M_x = 1$.

For defining the components of the vector $\mathbf{U}$ in subsonic part of the flow, the system of the equations can be written in the form:

$$\begin{pmatrix} U_2 \\ \frac{U_2^2 + \sigma E_1^2}{U_1} \\ \frac{U_2 U_3}{U_1} \\ \frac{U_2 U_4}{U_1} \\ \left(U_5 + (\gamma-1)\left(U_5 - \frac{U_2^2+U_3^2+U_4^2}{2U_1}\right)\right)\frac{U_2}{U_1} \end{pmatrix} = \mathbf{E}^{*i} \quad, \quad M_x = \frac{E_1}{\sqrt{\gamma p \rho}} \quad,$$

the solution of this system has the following form:

$$U_2 = E_1, \quad U_1 = \frac{E_1^2}{E_2}(1+\sigma), \qquad U_3 = \frac{E_3 E_1^2}{E_1 E_2}(1+\sigma),$$

$$U_4 = \frac{E_4 E_1^2}{E_1 E_2}(1+\sigma),$$

$$U_5 = \left(\frac{E_5 E_1 (1+\sigma)}{E_2} + \frac{\gamma - 1}{2(1+\sigma)} \frac{E_2^2 + E_3^2(1+\sigma)^2 + E_4^2(1+\sigma)^2}{E_2} \right) \frac{1}{\gamma}. \quad (8)$$

The second step is solved by the splitting method using the triple-step matrix sweep.

In order to suppress the high-frequency perturbations at the final step, the smoothing functions of the second and fourth orders are introduced into solution.

4 Calculation result

Numerical investigations were carried out for the following values of the characteristic parameters: $\gamma = 1.4$, $\mathrm{Pr} = 0.71$, $1 \le M_a \le 3$, $0.05 \le M_\infty \le 5$ and $1 \le n \le 10$. We use a grid with 101x101 nodes in the transverse directions with steps $\Delta y = \Delta z = 0.05$ while the step in the marching coordinate varied within the range $\Delta x = 0.0035 \div 0.015$.

The results of the calculations for the flow (the scheme of the flow is shown in Fig. 1a) are given for the outflow of the system of jets, which are located symmetrically about z-axis, into the cocurrent flow.

The initial data were specified at $x = 0.1$ for the condition $C_f = 1.5 \cdot 10^{-3}$, $T_w = 0.3$, $y^+ = 50$ and $Re = 2500$.

In order to reveal the features of underexpanded jets, the influence of the pressure ratio parameters on the flow pattern has been investigated. Figure 2A (a-d) ($M_a = 1.5$, $M_\infty = 2$, $n = 4$, $T_0 = T_\infty = 1$) shows the spatial pattern of the pressure field in the jet cross–sections $x = 4.54$, 13.7, 41, 77.5.

Clearly, in the process of propagation of the jet downstream, a shock wave begins to propagate from the initial section into the concurrent flow with a lower pressure (fig. 2A (a)). The wave reaches the boundary layer on the one side and the boundary of the calculation on the other side, reflects (fig. 2A (b)) and propagates in the opposite direction. In this case the reflection from the boundary layer occurs earlier than the reflection from the axis of symmetry of the jet (fig. 2A (c)). Thus, the initially axisymmetric shock wave after reflection from the initial boundaries becomes significantly three–dimensional. Then, as a consequence of intense dissipative processes caused by viscous and shock-wave processes, the pressure perturbation becomes weaker downstream (fig. 2A (d)).

The increase of Mach numbers of the cocurrent flow (fig. 2B, $M_a = 1.5$, $M_\infty = 5$) and the jet (the fig. 2C, $M_a = 3$, $M_\infty = 5$) leads to the appearance of perturbation waves, which moving from the wall towards the side of concurrent flow in both cases (fig. 2B, C (a)). From fig.2B, C (b) it follows that these waves subsequently interact with the shock wave, as a result a strong spike is observed (fig. 2B, C (c)). Then, this spike moves in a united front (fig. 2B, C (d)). As it follows from the figure, the velocity of propagation of the perturbation waves depends on Mach numbers. From the distribution of isobar in the plane xoy (fig. 3) for the above mentioned cases, it is obvious that at the increase of initial Mach numbers of the flow (fig. 3b, $M_a = 1.5$, $M_\infty = 5$) and the jet (the fig. 3c, $M_a = 3$, $M_\infty = 5$) at large distances from nozzles, a clearly expressed pattern of shock waves and waves of perturbation are observed, while at smaller Mach numbers (fig. 3a, $M_a = 1.5$, $M_\infty = 2$) the wave processes quickly weaken and the flow becomes isobaric.

Thus the outflow of the underexpanded jets from nozzle system into concurrent flow with partially limited domain becomes significantly three–dimensional because the velocity of the shock wave, reflected from the boundary layer, reaches the jet center faster than the wave, which is directed from the symmetry axis.

In order to investigate the features of the underexpanded supersonic jets extending in cocurrent subsonic flow, numerical experiments were carried out for $n = 1, 45 \div 4$. Numerical calculations are compared to the experimental data [9, 10], which were obtained for the investigations of jets flowing out in still air.

From the distribution of pressure in the plane xoy (fig. 4a) and along the axis x (fig. 4b) for $M_a = 2$, $M_\infty = 0, 25$, $n = 1, 45$, $T_0 = T_\infty = 1$, it follows, that the distinctive feature of underexpanded jets is that, in this case the structure of flow has periodic character which changes like a damping sinusoid curve. According to the general picture of flow of the supersonic jet into cocurrent subsonic flow at $p_0 > p_\infty$, the pencil of the characteristics (rarefaction waves) arises near the edge of nozzle, ensuring expansion of gas in the jet from pressure on of edge nozzle p_0 up to pressure of the environment p_∞. The waves of rarefaction reaches the boundaries of the jet and are reflected as a system of waves of compression.

Thus, a series of consistent waves of expansion and compression arise for outflow of underexpanded jets, therefore, the flow becomes periodic.

In fig. 4b the calculations (point–dashed lines) and experimental curve (dashed lines) of [10] are shown. Analyzing the results given in this figure, it is possible to note the quite acceptable correspondence of the numerical calculations of the present work with experiments, though the small divergence of results is observed depending on the distance from edge of nozzle.

At the decrease of the initial Mach number of the jet, up to the sound jet (fig. 5, $M_a = 1, 02$, $M_\infty = 0, 05$, $n = 1, 62$, $T_0 = T_\infty = 1$), the calculation results (continuous lines) are obtained, which are a well coordinated with the experimental data [4] (dashed lines), whereas the results of [10] (point-dashed line) show the divergences in the amplitude fluctuations already after fifth "barrel". From the distribution of pressure in a plane xoy (fig. 5b) a fast smoothing of the pressure of jet and the pressure of flow is noticeable.

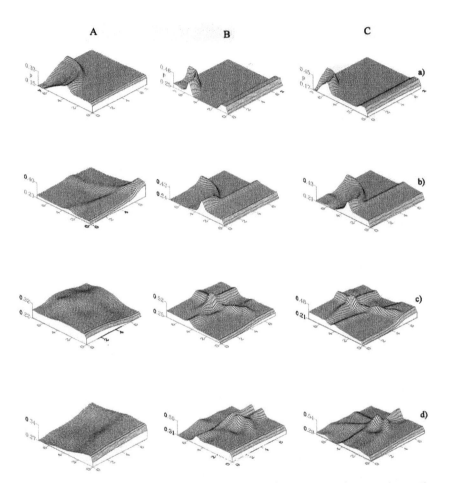

Fig. 2. Pressure profiles in the cross–sections: a) $x = 4.54$, b) 13.7, c) 41, d) 77.5; A. $M_a = 1.5$, $M_\infty = 2$, $n = 4$, $T_0 = T_\infty = 1$; B. $M_a = 1.5$, $M_\infty = 5$, $n = 4$, $T_0 = T_\infty = 1$; C. $M_a = 3$, $M_\infty = 5$, $n = 4$, $T_0 = T_\infty = 1$

Thus, from the above figures it can be seen that in numerical calculations of the present work and the works [10, 11], there are divergences with experimental data, which are not defined by the coincidence of phases in distributions of pressure on the axis of the jet.

a)   p

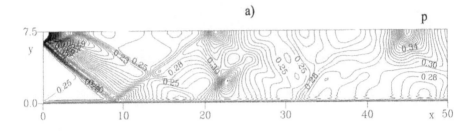

b)

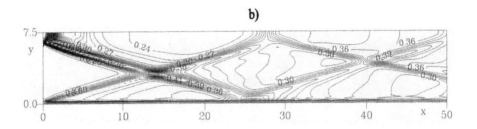

c)

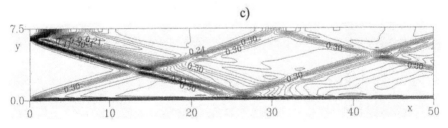

Fig. 3. Pressure field in the plane of xOy: a) $M_a = 1.5$, $M_\infty = 2$, $n = 4$, $T_0 = T_\infty = 1$; b) $M_a = 1.5$, $M_\infty = 5$, $n = 4$, $T_0 = T_\infty = 1$; c) $M_a = 3$, $M_\infty = 5$, $n = 4$, $T_0 = T_\infty = 1$

It is necessary to note, that in [10] the calculations were done by a two-parametrical model of turbulence, in [11] – a one-parametrical model of turbulence.

Thus, for correlation of calculation results and experimental data, there is the necessity for a further development of the model of turbulency allowing the calculations with $M_\infty = 0.0$.

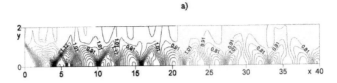

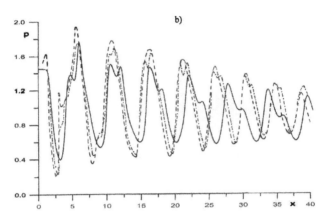

Fig. 4. a) Pressure field in the plane of xOy, b) variation of the pressure for $M_a = 2$, $M_\infty = 0.25$, $n = 1.45$, $T_0 = T_\infty = 1$

References

1. Peery KM, Forester CK (1980) AIAA J 9:1088–1093
2. Beam RM, Warming RF (1978) AIAA J 4:393–402
3. Kaltaev AZh, Naimanova AZh (2002) Math Model 14:105–116 (in Russian)
4. Fletcher CAJ (1988) Computational techniques for fluids dynamics. Springer Verlag, Berlin, Heidelberg
5. Loitsyanskii LG (1966) Mechanics of fluids and gases. Pergamon Press, Oxford
6. Schlichting H (1968) Boundary layer theory. McGraw-Hill, New York
7. Sheaff LB, Steger JL (1980) AIAA J 12:16–29
8. Anderson DA, Tannehill JC, Pletcher RH (1984) Computational fluid mechanics and heat transfer. McGraw-Hill, New York
9. Czyan Chzhesin (1967) Investigation of axisymmetric supersonic turbulent jet for the outflow from nozzle with underexpension. In: Abramovich GN (ed) Investigation of turbulent jets of air, plazma and real gas. Mashinostroenie, Moscow (in Russian)
10. Abdol-Hamid KS, Wilmoth RG (1989) AIAA J 3:315–322
11. Kozlov VE (1983) Method of calculation of weakly anisobaric underexpanded supersonic turbulent jet in subsonic cocurrent flow. In: Dulov VG (ed) Supersonic gas jets. Nauka, Novosibirsk (in Russian)

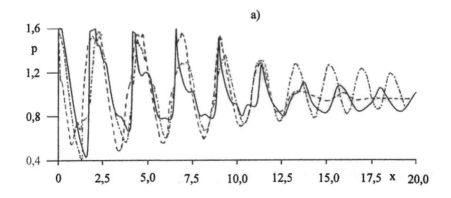

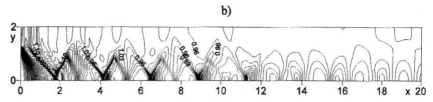

Fig. 5. a) Variation of the pressure, b) pressure field in the plane of xOy for $M_a = 1.02$, $M_\infty = 0.05$, $n = 1.62$, $T_0 = T_\infty = 1$

Object-oriented framework for parallel smoothed particle hydrodynamics simulations

S. Holtwick[1], S. Ganzenmüller[2], M. Hipp[2], S. Pinkenburg[2], W. Rosenstiel[2], and H. Ruder[1]

[1] Institute of Theoretical Astrophysics, University of Tübingen, Auf der Morgenstelle 10, 72076 Tübingen, Germany
`holtwick,ruder@tat.physik.uni-tuebingen.de`
[2] Department of Computer Engineering, University of Tübingen, Sand 13, 72076 Tübingen, Germany
`ganzenmu,hippm,pinkenbu,rosen@informtik.uni-tuebingen.de`

Summary. Smoothed Particle Hydrodynamics (SPH) is a widely spread method in scientific computing. It is a grid-free method for particle simulations. Most of the existing implementations are written in FORTRAN and C and therefore difficult to maintain and to extend. Here we describe the design and the implementation of a parallel object-oriented framework for particle simulations written in C++. The key features of sph2000 are easy configurability, good extensibility and the constantly expanding range of applications. The use of design patterns lead to an efficient and clear design, simplifying further algorithmic and methodical modifications and extensions. Advances made in the field of hybrid parallelization improve the efficiency and the portability of parallel applications. In addition the implementation of parallel I/O enhanced the performance significantly. The method was also upgraded by concepts to permit the simulation of compressible problems with free surfaces like spray atomization and to consider surface tension in these simulations.

1 Introduction

Within the Collaborative Research Center SFB382 "Methods and algorithms for the simulation of physical processes on supercomputers" by order of the DFG mathematicians, physicists and computer scientists cooperate in the field of computational physics to research new objectives in hydrodynamics and astrophysics. A new research project of the Landesstiftung Baden-Württemberg "Parallel simulation of intermixture, segregation and related instabilities with particle methods" follows up these workings to advance the SPH method and the concepts for its parallelization. Currently we apply SPH for problems in astrophysics, hydrodynamics, material sciences and granular media.

Resolution and accuracy of the simulations strongly depend on the total number of pseudo-particles and the average number of interactions between the particles. The current physical problems like the atomization of fluids need huge numbers of particles to obtain accurate results. The computational costs of SPH simulations are comparatively high. Therefore high-performance computers are indispensable. Although object-oriented programming became more common recently, there is still a lack of integration of object-oriented concepts for parallel scientific applications. Our group is making strong efforts to develop fast libraries for parallel particle simulations which are portable to many important platforms. Our intention is to provide a general framework for particle simulations based on the SPH method with reduced parallel overhead and minimized serial parts. This framework offers all advantages of object-oriented concepts. The libraries are clearly arranged, classes encapsulate their responsibilities and duties. Design patterns organize the areas of activity and decouple the classes. Uniform interfaces guarantee easy exchangeability and extensibility of the libraries. The sequential and parallel performance of sph2000 exceeds the performance of previous procedural implementations. As sample application we use simulations of diesel injection.

2 Related Work

To give an overview on important libraries for scientific computing the following workings have to be introduced. Cactus [1] is an extensive framework for parallel physical simulations with numerous modules, called thorns, which offer interfaces for different languages. POOMA [2] is a wide-spread framework for parallel physical computations as well. Both libraries, like most others, are designed for grid methods. Although POOMA supports moving particles, they have to be arranged in arrays. The main focus is on interactions and transformations between the particle arrays and the grid fields. Beside other SPH libraries, Gadget is a large library which offers standard algorithms for astrophysics with self-gravitation. Like most SPH libraries, it is not object-oriented. The embedding of pure particle methods in object-oriented libraries is unusual. To use the given object-oriented libraries, the developer of particle methods is obliged to take along the overhead of grid methods, deal with a higher complexity and comprehend the concepts of the library, although not every concept is needed. Using procedural SPH libraries means to abandon the advantages of object-oriented programming.

3 Method

Smoothed Particle Hydrodynamics is a grid-free particle method for the simulation of the compressible and viscous flow of gases and fluids. It was introduced in 1977 by Gingold and Monaghan [4] and primarily has become popular in astrophysics. The method provides a numerical algorithm to solve the systems of coupled partial differential equations for continuous field quantities. The formalism transfers such a system of equations to a system of coupled ordinary differential equations for values at discrete sampling points. The constitutive idea behind this particle method consists in the fact, that the sampling points, at which the system of equations gets evaluated, are moving according to the flow of the fluid. This corresponds to the lagrangian perception. Thus the real continuous fluid is represented by an ensemble of pseudo-particles. A certain SPH-particle represents the identical fraction of the real fluid during the entire course of the simulation considering all its thermodynamic and hydrodynamic properties and physical quantities.

As a result of this discretization the hydrodynamic base equations devolve into the equations of motion for an ensemble of interacting SPH-particles. Internal magnitudes such as density, pressure, temperature, internal energy and entropy are assigned to these particles. Some internal forces of the fluid, such as pressure and frictional tension, have to take effect on the SPH-particles as additional external forces. Hence the method is completely grid-free all interactions are computed by interpolation over the properties of the circumjacent particles. SPH, as a lagrangian method, is potentially self-adaptive because the allocation of the SPH-particles adjusts to the distribution of the mass-density.

The results of the simulations have got to be interpreted carefully under regard of certain constraints. The trajectories of the SPH-particles reveal the course of the physical flow and their frequency distribution represents the mass-density distribution of the fluid. During the interpretation of the particle distributions and movement is to consider that the individual SPH-particles interpenetrate each other almost arbitrarily. This circumstance necessitates specific concepts for the treatment of fixed boundaries within the simulated volume. The interpenetration of SPH-particles provokes that a border consisting of particles is not able to hold then up reliably. It is advisable not to assign any physical significance to the pseudo particles but simply to regard them as moving sampling points of the numerical method. The nonexistence of well defined and permanent relations of proximity between these individual sampling points proves to be a certain disadvantage of this particle method because it necessitates a computational expensive search for neighbours before each single time step and for each single SPH-particle. Since the method does not apply any grid applicability on complex geometries is guaranteed. The SPH algorithm is comparatively quick and simple to code.

The SPH formalism consists of three steps:

- a kernel smoothing, where all position dependent functions are being convoluted with a so called kernel function
- a Monte-Carlo integration, where the convolution integrals gained in the first step are being numerically evaluated and discretized

- the numerical integration of the resultant system of common differential equations concerning time

The fundamental idea is to smooth all spatial field quantities over a bounded volume by use of a convolution with a characteristic kernel function. The extension of this volume is determined by the smoothing length h. This kernel smoothing smoothes the virtual SPH-particles to a continuous density distribution and at the same time restricts the range of physical interactions. The kernel function has to be compact, monic and continuously differentiable. For the borderline case of a vanishing interaction length h the exact function is reproduced. Furthermore the kernel is spatially limited to keep the number of interactions finite, and normally the kernel is rotationally symmetric. The integrals are being discretized and thus transformed into sums. By use of partial integration all spatial deviations of the function f in the equations can be transferred to the kernel function W, whose deviations are analytically known. Thus from the system of coupled partial differential equations in space and time a system of common differential equations is attained. This system of equations can be integrated with conventional methods for solving common differential equations.

Because the SPH method is a lagrangian particle method, the time integration of the equations takes place at sampling points (SPH-particles) which are afflicted with mass and move along with the fluid. The applied method is the Monte-Carlo integration. All functions and integrals are evaluated only at the locations of the SPH-particles. An auxiliary fixed grid is not necessary. The method requires no fixed boundaries because the SPH-particles are freely movable. While in methods using grids the resolution is statically given by grid and exclusively depending on the lattice parameter in the SPH method the local resolution depends on the local SPH-particle density. At places with a high density of SPH-particles the system is computed with high resolution and calculation costs.

Several alternative formulations of the SPH formalism exist, which are applied on each equation of the system of differential equations. This leads to discretized differential equations in SPH description, which are used to build up the final numerical algorithm. In the simulation only by the discrete kernel smoothing approximated values of the simulation magnitudes are calculated. To apply the SPH formalism to a differential equation means to execute the kernel smoothing for the complete equation. The equation first is multiplied by the kernel function W and afterwards integrated over the entire volume V. The resultant equation is being transformed in such a way, that every single term describes an individual kernel smoothing. Finally the equation is discretized, whereas the summations are just executed for those terms which comprise spatial deviations. To complete the system of common differential equations gained under application of the SPH formalism information about the motion of the SPH-particles is needed. The common approach is to move the particles with a velocity proximal to the average velocity of the circumjacent flow. Some small adjustments are made with the intention that during the course of the simulation the SPH-particles retain a relative controlled regular composition. For flows which collide at high velocities this approach prevents the interpenetration of the particle flows which else would highly increase the variance of the results.

A very import parameter of SPH simulations is the selection of the kernel function. To restrict the computing costs the extension of the kernel has to be finite. Else the algorithm assumes an N^2 shape and each SPH-particle overlaps and interacts with every other particle. For the simulation of fluid jets the Cusp kernel has prevailed [5]. It owns the shape of a parabola sector, has a distinct apex in the centre and is as well steady differentiable even at the boundary. The kernel function and its deviations disappear at the outer boundary and the transition is steady and smooth. These properties favourably affect the evolution of the particle distribution, for instance it prevents the particles from lumping together and numerical effects such as oscillations are reduced. The possibility to choose from a variety of kernel functions makes the SPH method highly flexible.

A significant disadvantage of the SPH method compared to methods using grids are the by multiple times larger computing expense and memory require-ments. A large contingent of the calculating time is needed by the interaction search. The information on interactions, particularly the values of the ker-nel function and its deviations, are needed several times during the course of the calculations and therefore are kept in the main memory. Because every particle has between 100 and 200 interaction partners the demand on main memory for the interaction lists is enormous. Another reason for the huge computing expense results from the large ratio of particle speed and particle distance. Even for the fastest moving SPH-particle all interaction partners along its path have to be regarded. Each particle however is just able to see up to a certain horizon whose radius corresponds to the smoothing length. At a relative speed of the order of magnitude of one smoothing length per time step the interactions can not be accurately considered. Therefore at high res-olutions and large particle velocities the size of the time steps has to be very small to ensure the accurate computational consideration of all interactions.

4 Design and Implementation

Our main goal was to develop a parallel object-oriented SPH framework offer-ing extensibility, maintainability and reusability of the code. A main concern in the design was the strict decoupling of the parallelization on one hand from modelling aspects and physics on the other hand. Another goal was to prove the feasibility of an object-oriented approach in the performance crit-ical domain of particle simulations without loosing efficiency. The result is a parallel object-oriented particle simulation framework written in C++, called sph2000. Classes modelling the elements of the problem domain generate a well structured design. The use of several design-patterns helped to organize the classes clearly and efficient. They are introduced to structure the class library, to separate and group the application elements, as well as to define uniform interfaces. Additionally, they allow the insertion of extensions more simply because of decoupled elements. Table 1 shows the application elements and its corresponding design patterns.

Table 1. Application elements and corresponding design patterns

Element	Responsibility	Design Pattern
Initialization	Configuration, Object Creation	Builder, Configuration Table
Mathematics	Time Integration	Strategy, Iterator, Index Table
Physics	Particles, Interactions, Right Hand Side (RHS)	Strategy, Compositum, Iterator, Index Table, Proxy
Parallelization	Communication, Decomposition, Load Balancing	Strategy, Mediator
Geometry	Simulation Domain, Subdomains	Strategy, Decorator

The independent elements can be extended, causing no changes in other classes. The classes within an element can be easily exchanged because of uniform interfaces. Once implemented this enables the user to form new classes by simply copying and adapting the existing ones. Thus, extensions supplement the code instead of changing it.

Another goal was to simplify the configuration of simulations, which mainly means to configure a simulation run after the compilation at runtime by reading a parameter file. To avoid conditional compilation with preprocessor directives as it is often seen in C libraries, the `Strategy` pattern which is based on the object-oriented concept of polymorphism is used. With this pattern the program instantiates objects at runtime due to the configuration parameters. The complete configuration with all parameters of a simulation is stored in an object of the class `ParameterMap` (configuration table pattern). Every object which needs parameters owns a reference to the `ParameterMap` object. Thus all objects can access the configuration uncomplicatedly. Mainly the initialization objects (pattern `Builder`) access the `ParameterMap` to realize the exchangeability of the components. Every `Strategy` offers an accordant parameter to determine which concrete implementation is used in the simulation. In every simulation run, the `Builders` create and initialize only the needed objects. The concept of the configuration table makes the configuration data available for all auxiliary programs which work with simulation data. Such supplementary tools adopt the existing `ParameterMap` class without changes.

The flexibility of the framework is mainly up to the used physics. Because of enhancing the framework simultaneously with the physical method, the integration of additional physical quantities and the exchangeability of different calculation methods has to be guaranteed. An SPH-particle is a sampling point of the differential equations which moves with the flow and represents a volume element of the moving fluid. It contains all physical quantities of a fluid element, their interactions have to be evaluated and they have to be communicated among the subregions.

To protect the user from adapting the Particle class for every application, we introduced the class `IdStore` as an index table. The Particle class contains only the administrative structure for the communication by message passing and two STL containers for the scalar and the vectorial variables. The initialization determines the fixed number of needed variables within a simulation and writes them into the index table. Thus the particles can adapt themselves dynamically at start time to the respective simulation. The basis for this design is a set of two classes, the `QuantityBuilder` and the `IdStore`. The `QuantityBuilder` implements references to the used physics and initializes the `IdStore` object. It evaluates the parameters from the `ParameterMap` and reserves an index in the `IdStore` for every physical variable (e.g. position, speed and density). The physical variables are stored inside the particle in the order of these indices. Since the particle itself has no idea about the contained variables, classes needing a particles' variable have to get the information through the `IdStore`, which represents the interface to the variables.

The `QuantityBuilder` class is designed according to the `Builder` pattern. Besides the initialization of the `IdStore` it establishes the `QuantityList`, an STL container of calculation objects for the physical quantities. Since there are no general SPH formulas for the equations of motion, many different approaches evolved in the past. To achieve a high flexibility, the calculation objects are defined as a `Strategy` with a `Quantity` base class. The `QuantityBuilder` knows all possible quantities and their dependencies. By reading the physical parameters from the `ParameterMap`, quantities are selected and stored in the `QuantityList` in respect to the physical dependencies (see Figure 1). In each time step an RhsCalculator object iterates through the `QuantityList` to compute the right hand side (RHS) of the differential equations.

Like most elements of the framework, the Integrator class is also implemented as a `Strategy` and thus configurable and extendable like the quantity classes. The several Runge-Kutta and adaptive integrators of the framework are based on an abstract Integrator class. It defines interfaces to iterate through all particles' differential variables to integrate the right hand

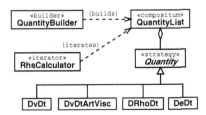

Fig. 1. Simplified class diagram of the sph2000 calculation classes

side and to store intermediate integration steps. The `Strategy` pattern is very easy to apply, since the user only has to write a configuration file, e.g. which `Integrator` should be utilized, and the `Builders` only create the needed objects of each `Strategy`.

5 Parallelization

The parallelization of SPH simulations is a challenge because of the very dynamic character of the method. Particles do not posses any static neighbourship relations causing a huge amount of communication between the computing nodes. New approaches for the parallelization of object-oriented codes on distributed memory architectures had to be devised.

For an efficient parallelization we implemented a mechanism for domain decomposition, dividing up the simulation area into several rectangular subdomains. These are equally spaced at the beginning of the simulation but dynamically change their size during runtime to keep the load balanced between all processors. The size is thereby given by the number of interaction partners since this linearly effects the calculation time. A class `SubSimulation` was implemented, which administrates each subdomain.

It is based on the design pattern `Strategy` to be able to differentiate between subdomains with different tasks. The framework knows two specialized types of `SubSimulations`: `SubSimulation`, `BoundarySimulation` and `InletSimulation`. `SubSimulation` defines the basic methods of a subdomain like administrating the geometry, the adjacent domains and the particles within the subdomain. `BoundarySimulation` extends this type by methods for reflecting and absorbing particles at the boundaries. `InletSimulation`, based on the `Decorator` pattern, can decorate the latter with additional methods for inserting new particles to the simulation through an inlet. The `ParameterMap` includes the initial and behavioural values of these three types.

In every time step each subdomain has to process the same tasks, like preparing the calculation, communicating particles to other subdomains, computing lists which contain the interaction partners, calculating the right hand side, or integrating the equations of motion. Each subdomain therefore contains classes and objects respectively for these tasks. From an object-oriented view each subdomain is a group of objects, which represent this area geometrically. All subdomains are independent from each other and communicate through a special communicator which is needed to pass objects from one subdomain to another. The implementation of this communicator follows the design patterns `Strategy` and `Mediator`. The communication and application flow within a group is implemented using an intra node `Mediator`, which knows all contained objects, coordinates the chronological processing of the tasks and uncouples all objects within a group from each other. To communicate particles and other information to and from the adjacent subdomains, an inter node `Mediator` is needed. This `Communicator` encapsulates the information about the

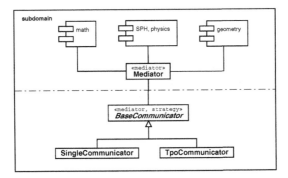

Fig. 2. Simplified diagram of a subdomain in sph2000. The upper part shows the modules with the SPH classes and the intra node mediator. The lower part shows the class diagram for the communication `Strategy`. The `Communicator` classes encapsulate the whole inter node communication. The calculation objects are completely decoupled from the parallelization

whole domain and communication structure. The implementation follows the design patterns `Strategy` and `Mediator`. The class `BaseCommunicator` defines the interface between intra node and inter node communication (see Figure 2).

The major advantage of this concept is that it enables the user to easily divide up the simulation domain into as many subdomains as processors are available. For communication between the processors the message passing object `TpoCommunicator` is generated, which uses TPO++, an object-oriented communication library set up on top of MPI. In case of a single processor simulation this communicator must be replaced by a single-node communicator. The user just has to exchange the communicators in the configuration file, leaving the code unchanged. To coordinate all subdomains a special master subdomain is implemented, which is extended by several administration objects. These are objects for timekeeping and administrating all particles as well as particle I/O objects for saving and restoring particle allocations.

6 TPO++

With MPI a widely accepted message passing standard has been developed. MPI provides support for object-oriented languages by defining C++ bindings for all MPI interface functions. But these are only wrapper functions which do not fit well into the object-oriented concepts. Moreover, MPI does not support the transmission of objects or standardized library data structures such as the Standard Template Library (STL) containers. To shorten the gap between the object-oriented software development and the procedural communication, we developed a communication library for parallelizing object-oriented applications called TPO++ [6] which offers the functionality of MPI 1.2. It includes a type-safe interface with a data-centric rather than a memory-block oriented view and concepts for inheritance of communication code for classes. Other goals were to provide a light-weight, efficient and thread-safe implementation and, since TPO++ is targeted to C++, the extensive use of all language features which help to simplify the interface. A distinguishing feature compared to other approaches is the tight integration of the STL. TPO++ is able to communicate STL containers and adheres to STL interface conventions.

7 Parallel I/O

The first results of sph2000 showed a significant lack in performance due to sequential I/O. Since current standards like MPI-2 only support procedural interfaces for parallel I/O, we extended TPO++ by an object-oriented interface for parallel I/O. The constantly growing gap between processor and hard disk performance during the last decades makes this problem even worse and necessitates the use of parallel I/O systems. The most often used interface for parallel I/O is the standard MPI-IO. One implementation of this standard is ROMIO, which is setup on top of the file system (e.g. XFS, PVFS, NFS, GFS) using an abstract device interface for I/O. However, MPI-2 only supports procedural interfaces for parallel I/O, making it impossible to be used in our object-oriented application. Other approaches like parallel HDF5 or NetCDF are also unusable, since they lack the integration of real object-oriented concepts. Therefore, we extended TPO++ by an object-oriented interface for parallel I/O. The major goals were to provide an efficient object-oriented interface, to provide the same functionality as MPI-IO, to be compliant to MPI-IO, and to support standard data types like Standard Template Library (STL) containers.

The initial version of sph2000 implemented a sequential I/O strategy. One master process gathered the part results from all other processes and saved the whole data in am ASCII format file to disk. The new strategy using the parallel I/O interface was to implement collective I/O. Thereby, all processes can access the same file in parallel, which improves the performance significantly, since the communication to the master process can be omitted and the whole parallel I/O bandwidth can be used for transferring the data to and from the disks. In addition, the library internally calculates the correct offsets within the file where each process has to place its part, avoiding any extra implementation by the user. To provide sph2000 with parallel I/O, the particle I/O object had to be adapted by Pinkenburg [7]. The following listing shows the adapted method `saveDataFile` and represents the simple usage of the interface:

```
#include<tpo++.H>

void ParticleIO::saveDataFile(const ParticleContainer\& particles,
                                                    string name)
{
```

```
TPO::File fh;
int code = fh.open(TPO::CommWorld, name, TPO_MODE_CREATE);
fh.write_all(particles.begin(), particles.end());
fh.close();
}
```

This implementation of using a single collective call (`fh.write_all`) reduces the size of the original code by about 100 lines of code. The call is needed to save the containers of particle objects of each processor to disk simultaneously. Two iterators called `begin()` and `end()` thereby define the beginning and ending of the container. This syntax also enables the user to store only a subset of particle objects to disk. The performance improvement through the usage of parallel I/O within the framework depends on the application as well as on the ratio of computation and I/O within the application. To determine the real performance gain we implemented a sample application.

8 Hybrid Parallelization

There is a general need for larger simulations needing more memory and computing power. The majority of parallel machines in the TOP500 list are hybrid parallel architectures, a combination of N-way shared memory nodes with message-passing communication between the nodes. On an increasing number of hybrid architectures a parallelization combining threads and message-passing is a promising way to reduce the parallel overhead in respect of memory and performance compared to pure message-passing parallelization. Hybrid parallelization is the combination of a thread based programming model for parallelization on the shared memory nodes with message-passing based parallelization between the nodes. The standard libraries are OpenMP for thread based programming and MPI for message passing.

The sharing of common data structures on shared memory nodes reduces the amount of required memory. Even more important is the reduction of communication. The maximum speedup of a parallel implementation is limited by its serial parts. In our non trivial particle codes the major serial part is the (often collective) communication. Hybrid parallelization reduces the amount of transferred data because the communication of shared data on SMP nodes is implicit. But the communication itself is faster, too. A simple test shows this for the MPI-Allgather call. We distribute a 200 MB data array on the Hitachi SR8000 between all nodes. K is the number of nodes. In the first test (pure-MPI) a MPI process runs on every SR8000 processor. The $K \times 8$ processors send and receive $200/(K \times 8)$ MByte to/from each *processor*. In the second test (Hybrid) the same amount of data is just sent between the master threads of each node and thus each call sends and receives $200/K$ MByte to/from each *node* (other non-master threads are idle). The values in table 2 give the time for 50 MPI-Allgather calls in seconds on the SR8000-F1 at the HLRB in Munich. One can see a big difference between the hybrid communication (comparable to the hybrid programming model with implicit intra node communication) and the pure-MPI communication model with explicit MPI intra node communication. For the $K = 32$ run, the hybrid programming model is 8 times faster. Hybrid parallelization is portable and performant on a variety of parallel architectures with shared and distributed memory. As Hipp [8] shows Especially for large node numbers hybrid parallelization shows significant performance advantages compared to pure message-passing.

Table 2. Time for 50 MPI-Allgather calls in seconds on Hitachi SR8000-F1

	K=1	K=2	K=4	K=8	K=16	K=32
Pure-MPI	16	51	97	126	172	200
Hybrid	10	15	20	21	22	25

9 Sample Application

The general intention of our research is to develop a cumulative framework with simulation methods and evaluation procedures which allow the prediction of all fundamental technical values in unsteady fluid atomization and spray processes. Our main focus is lying on the primary jet breakup of Newtonian fluids. Important questions are the velocity fields of the fluid and the gas and their time dependency, instabilities, surface tension, density and viscosity of the surrounding gas, turbulence, activities on the boundary layer and cavitation with its implication on the atomization. Evaporation might be a long-term problem. We collaborate with working groups who do the experimental characterization of the jet breakup to verify the simulation results. We contribute to the modelling of the constriction of fluid ligaments and the breakup of compact fluid jets. In the future the attained insight might be used to optimize the geometry of nozzles to obtain sprays with improved drop size distributions.

Approaches made for the simulation of free surfaces were already introduced in 1994 by Monaghan [9]. The SPH method is in principle able to describe the breakup of diesel jets. Elementary investigations on this application were described by Ott [10]. Examinations on the influence of inflow boundary conditions on the jet breakup at the beginning of the injection revealed that a realistic breakup behaviour can only be achieved by stochastic disturbance at the inlet. The viscous stress tensor was implemented in the SPH formalism to be able to simulate the complete Navier-Stokes equations. Ott has enhanced the SPH method to be able to simulate flows of two separated phases such as occur in the primary jet breakup. This enhancement has been examined by several test problems and proved stable even for huge density variations. Since the SPH method is able to simulate compressible flows the method is highly applicative to simulate the processes at extremely high injection pressures.

The major difficulty in investigating the atomization process of a compact fluid jet close to the nozzle is the huge particle density within the spray. As far as we know there is not yet any physically consistent and experimentally verified concept for the mechanism of atomization under conditions comparable to an engine. A theoretical satisfactory approach is complicated by disturbances generated in the nozzle. Therefore we concentrate on pure fluids and nozzles generating a spray which is as undisturbed as possible. Pressure is an important parameter for the atomization of fluids. Diesel engines with direct injection for the fuel-mixture generation produce pressures up to 1500 bar. The appearing phenomena like turbulence of the jet and cavitation do not only depend on the injection pressure but as well on the density in the chamber and the geometry of the nozzle. For a better understanding of the proceedings in the area close behind the nozzle the disintegration process is investigated numerically and experimentally. Important quantities to simulate and to verify are the structure of the fluid jet, the speed of the jet proximate to the nozzle emission and the velocity field of the surrounding gas.

For the simulation of the breakup we use Smoothed Particle Hydrodynamics. The main advantage compared to many other methods is the possibility to simulate compressible processes. The effects of the aerodynamic interactions are calculated directly. By comparison with a non compressible Volume of Fluid method (FLUENT) applied by our colleagues in Aachen the potential of both methods are acquired. For most of our simulations we are using a self-made Linux cluster based on standard PC components equipped with a high performance Myrinet network. Other hardware is made available by the HLRS in Stuttgart and the HLRB in Munich.

10 Results

Extensive parameter studies approved the correctness of the implementation. So far the sph2000 framework provides 5 kernel functions, 6 integrators, and 20 quantities to calculate the state equations and the equations of motion for the air and diesel particles. 2D and 3D simulations with up to 2 million particles reveal a broadening and breakup of the diesel jet leading to turbulences behind the dispartment. After short time single drops are separated from the compact jet, see Figure 3.

The left figure illustrates the distribution of the sampling points after 1600 ns at an injection speed of 400 $\frac{m}{s}$. Diesel particles are shown blue, air particles are shown red. Every individual sampling point is displayed. The regular patterns in both domains are not numerical effects but beats of the particle positions with the limited resolution of the illustration. The right figure shows the density distribution. The injected diesel is shown dark because of its huge density compared to the air. The front of the jet is deformed and broadened. First drops are split off from the edge of the jet. The grey spherical shape displays the density wave caused by the injected diesel. The speed of the jet approximates 60 percent of the acoustic velocity in this configuration. Therefore the density wave is anticipating in front of the jet.

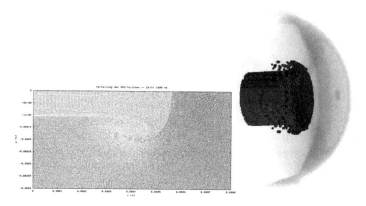

Fig. 3. 2D and 3D simulation runs with 1 million particles

The examination of effects at the order of magnitude of the smoothing length is problematic. The SPH-particles adjoin much closer and it does not make sense to evaluate the field quantities for every single particle. Therefore the field quantities are interpolated on a cartesian grid whose lattice parameter is of the same order of magnitude as the smoothing length.

The performance of the libraries has been measured on our Kepler-Cluster, a self-made clustered supercomputer based on commodity hardware. Its architecture consists of two parts: The first diskless part with 96 nodes each running with two Pentium III processors at 650 MHz and having 1 GB of total memory, equaling 512 MB per processor, and the newer second part with 32 nodes each running with two AMD Athlon processors at 1.667 GHz, sharing 2 GB of total memory, and using an 80 GB disk. The whole system has two interconnects, a fast Ethernet for booting the nodes, administration purposes and as storage area network (SAN), and a Myrinet network for the communication between parallel applications. Both have a multi-staged hierarchical switched topology organized as a fat tree. The Myrinet has a nominal bandwidth of 133 MB/s which is the maximum PCI transfer rate. The parallel I/O architecture consists of the MPI-IO implementation ROMIO, which is set up on top of the file system using an abstract device interface for I/O (ADIO). The underlying parallel file system is PVFS which is configured as follows: The 32 disks of the AMD nodes act as I/O nodes and the whole 128 nodes are PVFS clients. The SAN is used for striping the data onto these disks.

We compared the efficiency of our library on this platform using simple read/write tests as well as collective read/write tests and measured the achieved aggregated bandwidths of MPI and TPO++. Figure 4 shows a set of our measurements on Kepler. The overall limit is determined by the 32 I/O nodes using PVFS over the SAN. The I/O nodes are divided into two parts connected hierarchically to a Fast Ethernet backplane. Measurements give an effective aggregated bandwidth of about 180 MB/s. The differences seen in all charts between MPI and TPO++ represent the loss in performance due to the object-oriented abstraction. Evidently, object-oriented parallel I/O using TPO++ achieves almost the same bandwidth as MPI-IO.

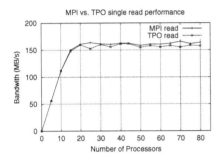 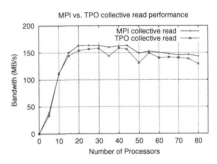

Fig. 4. Comparison of parallel I/O using MPI and TPO++

Furthermore we measured two different simulation setups: The first running on the Pentium nodes with disabled I/O and the second on the Athlons with I/O enabled in every second time step, to compare the performance of sequential and parallel I/O. We always used only one processor per node.

Table 3 shows the performance results of setup 1. Due to memory shortage of the Pentium nodes, the measurements were made starting with 2 nodes. The parallelization scales very good up to 64 processors leading to a remarkable speedup of 44.10.

Table 3. Results of the first simulation setup with 1 million particles on Pentium

Processors	1	2	4	8	16	24	32	64
Time per step (in s)	-	116.9	58.8	33.8	16.4	11.3	9.7	5.3
Speedup	-	2.00	3.98	6.90	14.24	20.68	24.10	44.10

The results of setup 2 show the significant effect of parallel I/O on the performance (see Table 4). The I/O part could be improved by a factor of 20 when using 32 processors working on a parallel file system with 32 distributed disks. Since I/O is only a small proportion of the whole simulation, the overall gain using parallel I/O reduces to a factor of 3.5 (64 processors). Note that even the sequential simulation with parallel I/O is faster than without parallel I/O, due to changing from ASCII to binary format and less code overhead for saving the particles.

Table 4. Results of the second simulation setup with 1 million particles on Athlon

Processors	1	2	4	8	16	24	32	64
Execution time (in s)								
- sequential I/O	2090.2	1050.9	530.4	407.0	349.4	305.1	272.4	251.6
- parallel I/O	1223.5	617.2	312.5	191.4	130.9	96.5	79.8	71.3
Speedup								
- sequential I/O	1	1.98	3.94	5.13	5.98	6.85	7.67	8.30
- parallel I/O	1	1.98	3.91	6.39	9.34	12.67	15.33	17.15

11 Conclusion

The application of object-oriented development methods has improved the quality of our simulation codes. The implementation is very easy to maintain and to extend, e.g. to modify the physics. The result of object-oriented techniques with design patterns is a framework, in which classes have clear and strictly separated responsibilities. Different methods can be interchanged without influencing other code. The use of our parallel I/O library with its optimizations for communication reduced the sequential parts of the framework to a minimum. These workings lead to a well scaling parallel performance.

In the future, we focus on the development of models for simulating cavitation and turbulence. These extensions will again require a strongly increased number of particles to simulate these effects meaningfully. Therefore we are investigating new solutions to decrease the amount of calculated interactions to reduce the runtime of the simulations. It is not sufficient only to optimize the communication. We furthermore try to avoid calculations with minor importance. A first approach is to prevent the density wave triggered by the injected diesel to be reflected at the outer boundaries. Another concept is to generate the majority of the air particles at runtime during the course of the simulation as soon as they are required. The basic idea is to calculate only air regions which are affected by the diesel jet and put in motion by the density wave. A great advance would be the employment of integrators that allow variable time steps for individual particles. This would minimize the computing costs for large areas with few dynamic in the simulation. The parallelization of the given extensions will require new concepts. Especially the load balancing mechanisms will have to be refined.

References

1. Allen G, Goodale T, Seidel E (1999) The Cactus Computational Collaboratory: Enabling technologies for relativistic astrophysics, and a toolkit for solving pdes by communities in science and engineering. In: Proc. of the 7th Symp. on the Frontiers of Massively Parallel Computation – Frontiers 99. IEEE Computer Society
2. Reynders JVW, Cummings JC, Hinker PJ, Tholburn M, Banerjee MSS, Karmesin S, Atlas S, Keahey K, Humphrey WF (1996) Pooma: A framework for scientific computing applications on parallel architectures. In: Wilson GV ,Lu P (eds) Parallel Programming using C++. MIT Press
3. Springel V, Yoshida N, White SDM (2001) New Astronomy 6(3):51
4. Gingold RA, Monaghan JJ (1977) Mon Not R Astr Soc 181:375–389
5. Speith R (1998) Analysis of SPH on the basis of astrophysical examples. PhD Thesis, University of Tübingen, Tübingen (in German)
6. Grundmann T, Ritt M, Rosenstiel W (2000) TPO++: An object-oriented message-passing library in C++. In: Proc. of the 2000 International Conf. on Parallel Processing. IEEE Computer Society
7. Pinkenburg S, Rosenstiel W (2004) Parallel I/O in an Object-Oriented Message-Passing Library. In: Kranzlmuller D, Kacsuk P, Dongarra JJ (eds) Recent Advances in Parallel Virtual Machine and Message Passing Interface: Proc. of the 11th European PVM/MPI Users' Group Meeting. Lecture Notes in Computer Science 3241. Springer Verlag
8. Hipp M, Rosenstiel W (2004) Parallel Hybrid Particle Simulations using MPI and OpenMP. In: Danelutto M, Vanneschi M, Laforenza D (eds) Euro-Par 2004 Parallel Processing: Proc. of 10th International Euro-Par Conf. Lecture Notes in Computer Science 3149. Springer Verlag
9. Gingold R A, Monaghan J J (1994) J Comput Phys 110(2):399–406
10. Ott F (1999) Advancement and analysis of SPH to simulate the breakup of free jets in air. PhD Thesis, University of Tübingen, Tübingen (in German)
11. Holtwick S Simulation of diesel injection using SPH on the Linux-Cluster Kepler. Diploma thesis, University of Tübingen, Tübingen (in German)

Numerical calculation of industrial problems

U.K. Zhapbasbayev, G.I. Ramazanova, and K.B. Rakhmetova

al-Farabi Kazakh National University, Al-Farabi av. 71, 050078 Almaty,
Kazakhstan `nich7@kazsu.kz`

Summary. The results of research of industrial problems, such as, pipeline transport of oil and reforming in catalytic reactor are presented in the paper. The first
part of the work is devoted to the calculation of hydrodynamics and heat exchange
of a turbulent pipe flow of oil-mixture with additives. In the second part the mathematical model of reforming process and results of calculation of aerodynamics, heat
and mass transfer in a catalytic reactor are presented.

1 Calculation of hydrodynamics and heat exchange of depressant technology of oil transportation

The addition of an extremely small amount of some high-molecular polymers
to a turbulent fluid flow leads to reduction of turbulent friction and heat
transfer. Many experimental and numerical studies of drag reduction mechanism by polymer additives in turbulent pipe flow were carried [1]-[2]. Thus a
drag-reduction rate of 46.8% and heat transfer reduction rate of 54.5% were
obtained.

The growth of volumes of paraffinic oil transportation on pipelines requires increasing of efficiency of oil pipeline operation. For increase of pipeline
throughput it is necessary to decrease of turbulent shear stress. One of effective ways of reduction of hydrodynamical resistance of turbulent pipe flow is
using of the anti-turbulent additives [2]-[4].

In this connection carrying out of calculations of hydrodynamics and heat
transfer of paraffinic oil transportation on pipelines is necessary.

We consider a stabilized turbulent oil flow in a pipe with length L and
radius R_w. Movement of oil is considered in cylindrical coordinate system,
the axis OZ is directed on the pipe axes and the axis OR - on the pipe radius.
Because of axial symmetry of the problem to axis OZ the area $0 \leq z \leq L$ is
considered (see fig.1).

The additive is entered in near wall layer of the flow. Concentration of
the additive can change in the range 0 - 500 ppm. As the concentration is

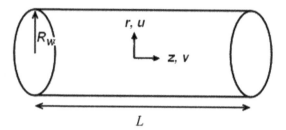

Fig. 1. The scheme of flow

very small it is considered, that the additive operates in wall zone, reducing friction on a wall.

The action of the additive on hydraulic resistance of water solutions can be presented in the universal relation form. By analogy it is considered, that for a oil-mixture flow with additives the ration of hydraulic resistance in a pipe can be presented in the form:

$$\frac{\zeta}{\zeta_o} = 1 - \phi(c), \tag{1}$$

where ζ, ζ_o are hydraulic resistance coefficients of a pipe flow with effect of additive and without one, accordingly, c - concentration of an additive.

Dependence of relative decrease of hydraulic resistance on additive concentration is presented in figure 2. With increase in concentration it is observes sharp decrease of resistance and at some value of concentration it passes to asymptotic.

The function $\phi(c)$ is well approximated by the formula:

$$\phi(c) = a * arctg\left(\frac{\pi}{b}c\right),$$

where a and b are the constants equal to $a = 0.375$, $b = 15.2$.

As a result of generalization of experimental data universal relation between hydraulic resistance of a oil-mixture pipe flow with adding additive and without one is established. Using (1), it is easy to establish connection between friction velocities of Newtonian flow and flow with additive effect.

Expressing hydraulic resistance in friction velocity we receive the formula:

$$\frac{v_*}{v_*^o} = \sqrt{1 - \phi(c)}, \tag{2}$$

where v_*, v_*^o are friction velocities of Newtonian flow and flow with additive effect.

The expression of friction velocity (2) allows to use (k - ε) model of turbulence for calculation of turbulent flow and heat exchange of a oil-mixture at influence of an additive.

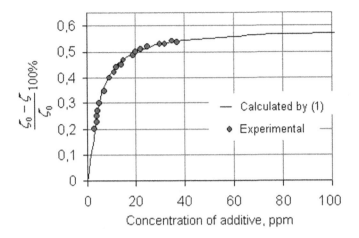

Fig. 2. Maximum decrease in hydraulic resistance versus concentration of depressant additive

Non-isothermal oil pipe is described by the system Reynolds-averaged Navier-Stocks equations [7]-[9]:

$$\rho u \frac{\partial u}{\partial z} + \rho v \frac{\partial u}{\partial r} = -\frac{\partial p}{\partial z} + 2\frac{\partial}{\partial z}\left(\mu_\varepsilon \frac{\partial u}{\partial z}\right) - \frac{2}{3}\frac{\partial}{\partial z}\left(\mu_\varepsilon div\overrightarrow{V}\right) +$$
$$+ \frac{1}{r}\frac{\partial}{\partial r}\left[\mu_\varepsilon r\left(\frac{\partial u}{\partial r} + \frac{\partial v}{\partial z}\right)\right], \tag{3}$$

$$\rho u \frac{\partial v}{\partial z} + \rho v \frac{\partial v}{\partial r} = -\frac{\partial p}{\partial r} + \frac{2}{r}\frac{\partial}{\partial r}\left(\mu_\varepsilon r\frac{\partial v}{\partial r}\right) - \frac{2}{3}\frac{\partial}{\partial r}\left(\mu_\varepsilon div\overrightarrow{V}\right) +$$
$$+ \frac{\partial}{\partial z}\left[\mu_\varepsilon\left(\frac{\partial u}{\partial r} + \frac{\partial v}{\partial z}\right)\right] - \frac{2v}{r^2}\mu_\varepsilon, \tag{4}$$

$$\frac{\partial \rho u}{\partial z} + \frac{1}{r}\frac{\partial(\rho v r)}{\partial r} = 0. \tag{5}$$

The equation of heat transfer can be written as:

$$\rho c_p \left(u\frac{\partial T}{\partial z} + v\frac{\partial T}{\partial r}\right) = \frac{\partial}{\partial z}\left(\lambda_\varepsilon \frac{\partial T}{\partial z}\right) + \frac{1}{r}\frac{\partial}{\partial r}\left(r\lambda_\varepsilon \frac{\partial T}{\partial r}\right). \tag{6}$$

In the equations (3) – (6): z, r – cylindrical coordinates; u, v – components of speed vector $\overrightarrow{V}$; p, ρ, T, c_p- pressure, density, temperature and thermal capacity of oil; $\mu_\varepsilon = \mu + \mu_t$, μ - dynamic viscosity of coefficient of oil, μ_t- factor of turbulent vortical viscosity; $\lambda_\varepsilon = \lambda + \lambda_t$, λ- factor of heat conductivity; $\lambda_t = c_p\mu_t/\mathrm{Pr}_t$, Pr_t - turbulent Prandtl number. Prandtl number Pr is defined on initial temperature T_0 and is accepted equals to $\mathrm{Pr}=0.9$. The turbulent analogue of number Prandtl Pr_t according to [10], [11] can be taken as $\mathrm{Pr}_t=0.9$.

Thermal properties of oil have been found according to experimental researches of oil with depressant additives and generalized in the form of following empirical dependencies:

$$\rho = 18.5 \cdot \exp\left(-\frac{T - T_o}{72}\right) + 821.5 \ (kg/m^3);$$

$$\mu = 5.511 \cdot \exp\left(-0.1487 \cdot T\right) + 0.005853 \ (kg/(m \cdot s));$$

$$\lambda = 0.042 \cdot \exp(-0.006695 \cdot T) + 0.124 \ Wt/(m \cdot grad).$$

In the temperature range $303K \leq T \leq 373K$ a thermal capacity of oil c_p slightly varies and remains to a constant $c_p{=}0.23$ kJ / (kg·grad).

Turbulent dynamic viscosity coefficient μ_t is defined on the basis of (k - ε) models of turbulence intended for low numbers of Reynolds Re, and expressed by the formula [11]:

$$\mu_t = C_\mu \rho \frac{k^2}{\varepsilon} f_\mu, \tag{7}$$

where ε – dissipation rate of kinetic energy of turbulence, f_μ – wall function [11]: $f_\mu = 1 - \exp(-0.0115y_+)$.

The differential equations for kinetic energy of turbulence and its dissipation rate are given by

$$\rho u \frac{\partial k}{\partial z} + \rho v \frac{\partial k}{\partial r} = \frac{\partial}{\partial z}\left(\left(\mu + \frac{\mu_t}{\sigma_k}\right)\frac{\partial k}{\partial z}\right) + \frac{1}{r}\frac{\partial}{\partial r}\left(r\left(\mu + \frac{\mu_t}{\sigma_k}\right)\frac{\partial k}{\partial r}\right) + \\ +P_k - \rho\varepsilon - D, \tag{8}$$

$$\rho u \frac{\partial \varepsilon}{\partial z} + \rho v \frac{\partial \varepsilon}{\partial r} = \frac{\partial}{\partial z}\left(\left(\mu + \frac{\mu_t}{\sigma_\varepsilon}\right)\frac{\partial \varepsilon}{\partial z}\right) + \frac{1}{r}\frac{\partial}{\partial r}\left(r\left(\mu + \frac{\mu_t}{\sigma_\varepsilon}\right)\frac{\partial \varepsilon}{\partial r}\right) + \\ +C_1 \frac{\varepsilon}{k} P_k - C_2 \frac{\rho\varepsilon^2}{k} f_e - E, \tag{9}$$

where f_e – wall function, and D, E – the functions expressing wall boundary conditions for k and ε.

In the case of additive influence friction velocity of flow is defined by the formula (2) and is used for definition the wall functions entering in the (k - ε) models of turbulence. Friction velocity v_*^o for Newtonian oil flow is calculated by (k - ε) models of turbulence.

The heat transfer equation is written concerning excess temperature $\theta = (T - T_w)/(T_o - T_w)$, where T_w - environment temperature.

The system of the equations (3) - (9) is led to dimensionless form. Coordinates z, r are divided on radius of a pipe, velocity components of u, v – on mean velocity, pressure p – on the maximal value of a dynamic pressure, temperature T – on T_0, density, dynamic viscosity, heat conductivity and thermal capacity - on values of these magnitude at temperature T_0.

Boundary conditions of the problem:

$$at \ r = 0: \ 0 \leq z \leq L; \ \frac{\partial u}{\partial r} = v = \frac{\partial \theta}{\partial r} = \frac{\partial k}{\partial r} = \frac{\partial \varepsilon}{\partial r} = 0;$$

$$at \ r = R_w: \ 0 \leq z \leq L; \ u = v = k = \varepsilon = 0; \ -\frac{\partial \theta}{\partial r} = Bio\theta. \tag{10}$$

The system of the equations (3) - (9) with boundary conditions (10) is solved numerical method in stream function ψ and vorticity ω variables. The differential equation of stream function ψ is discretised using the over relaxation scheme. The differential equation of vorticity ω is approximated using the up stream scheme of the second type. The Gauss-Seidel method is used to solution of vorticity, heat transfer and turbulent kinetic energy equations.

Some version of (k - ε) models, which differ by expression of wall functions, boundary conditions for k and ε on a wall and value of constants, are used (see tab. 1-3).

Table 1. Constants in the turbulence models

Model	Code	C_μ	T_1	T_2	σ_k	σ_ε
Chien (1982)	CH	0.09	1.35	1.80	1.0	1.3
Myong-Kasagi (1990)	MK	0.09	1.40	1.80	1.4	1.3
Low-Reynolds-Number (2003)	LRN	0.09	1.40	1.80	1.4	1.3

Table 2. Wall functions

Code	f_μ	f_e
CH-MK	$1 - \exp(-0,0115 y_+)$	$1 - 0,22\exp(-Re_t^2/36)$
MK-LRN	$[1 - \exp(-y_+/70)]\times$ $\times\ 1 + 3,45/Re_t^{1/2}$	$\left\{1 - \frac{2}{9}\exp\left[-\frac{Re_t^2}{36}\right]\right\}\times$ $\times\ 1 - \exp\left(-\frac{y_+}{5}\right)^2$

Table 3. D and E terms and wall boundary conditions for k and ε

Code	D	E	Boundary conditions
CH-MK	0	0	$k_w = 0$, $\varepsilon_w = v\frac{\partial^2 k}{\partial y^2}$
MK-LRN	0	0	$k_w = 0$, $\varepsilon_w = v\frac{\partial^2 k}{\partial y^2}$
Note	$Re_t = \frac{k^2}{v\varepsilon}$, $y_+ = \frac{yV_*}{v}$, $V_* = \sqrt{\tau_w/\rho}$		

For verification of the models and methods of the numerical solutions of the system of equations the test problem of calculation of the developed turbulent Newtonian pipe flow has been considered. Results of numerical calculation of the test problem have shown, that the combination models of Chen [11] and Myong-Kasagi [12] (CH-MK), Myong-Kasagi and LRN – models [11] (MK – LRN) qualitatively well describe distribution of kinetic energy of turbulence

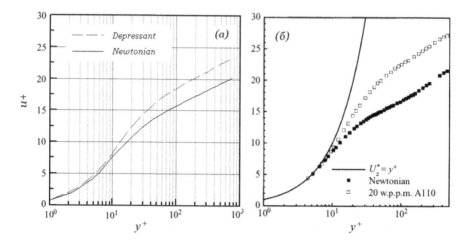

Fig. 3. Dimensionless velocity profiles

in a zone of the greatest generation of turbulence energy. In this connection calculations of turbulent flow and heat exchange with an additive have been lead by the indicated models of turbulence.

Calculations have been carried out at Reynolds numbers Re: 10550, 25000, 50000 and values of concentration of additive FLO-XL 10 – 100 ppm.

Figure 3 (a) shows profiles of axial velocity on logarithmic coordinates at value of Reynolds number Re=10550 and concentration of the additive c=20ppm and figure 3 (b) shows experimental data [1]. From figure 3 it is visible, that in the case of Newtonian flow the well-known logarithmic law of velocity distribution is carried out. At influence of the additive distribution of axial velocity in a viscous sublayer coincides and, since its external border, starts to increase and at $y^+ > 30$ axial velocity profile passes to in parallel the logarithmic law of pipe. It turns out, that in the case of additive solution flow the profiles of axial velocity is extended, i.e. turbulent transfer decreases in a radial direction, and it raises in a longitudinal direction. Distributions of axial velocity for additive solution flow and Newtonian flow are in the good consent with experimental data [1] (see fig. 3 b).

The distribution of turbulent shear stress near wall characterizes resistance of turbulent pipe flow. In this connection turbulent shear stress in a radial direction have been calculated. This is illustrated by figure 4.

As it can be seen in figure 4(a) the peak of the turbulent shear stress profile is shifted away from the wall to a higher $y+$ value, and the magnitude of the peak is decreased at wall zone.

In figure 4 (b) experimental data of turbulent shear stress profiles are presented. There is decreasing of turbulent shear stress at wall zone where the greatest generation of kinetic energy of turbulence, which is conformed by

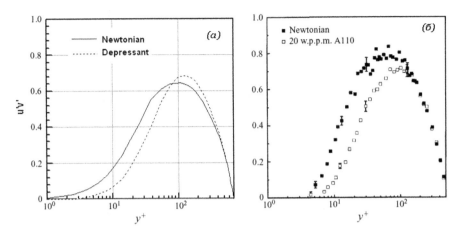

Fig. 4. Profiles of turbulent shear stress:(a) – data of numerical calculation; (b) – experimental data [1]

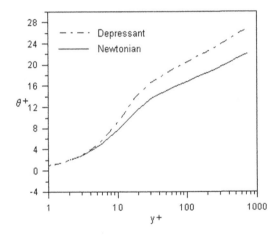

Fig. 5. Dimensionless temperature profile

the experimental data. The calculation data of turbulent shear stress are in the satisfactory agreement with the experimental data.

Figure 5 shows temperature profiles in logarithmic coordinates at values of Reynolds number Re = 10550 and additive concentration c=20 ppm. We shall notice that distribution of temperature at presence of additive effect lays above in a turbulent part of flow, also as the longitudinal velocity profile. It is possible to consider, that turbulent transfer to a radial direction decreases at presence of an additive, and, on the contrary, it raises in longitudinal direction.

In fig. 6 the profiles of radial components of a turbulent heat flow $-v'\theta'$, describing distribution of temperature are shown at Re = 10550 and c=20ppm.

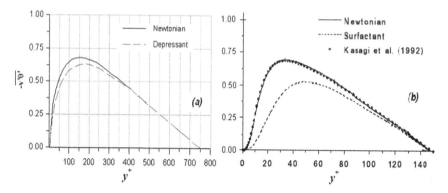

Fig. 6. Profiles of turbulent heat flux(a) – data of numerical calculation; (b) – experimental data [2].

Apparently from figure 6 (a), settlement distribution - $v'\theta'$ at presence of an additive lays below, than without its influence. Such behaviour of the - $v'\theta'$ profile at presence of an additive well proves to be true in experiences [2]. Thus, it is possible to consider, that turbulent heat transfer on radial direction decreases at input of additive in.

In all considered cases the received calculation results are in the satisfactory consent with experimental data and show possibility of application of k-ε model of turbulence for the description of flow and heat exchange at effect of additives.

As a result of generalization of the pilot tests of depressant additives on sites of main oil pipelines are received criteria dependencies of laws of resistance and heat exchange of oil-mixture flow with additive.

The received logarithmic profiles of velocity the basis of experimental data looks like:

$$\frac{u}{v_*} = 7.1 \lg \frac{y\,v_*}{\nu} - 8.5. \tag{11}$$

The dependence of resistance factor on Reynolds number for oil with depressant additive passes between the law of resistance for smooth pipes and Virk's asymptotic (for maximal decrease in resistance (fig.7)), and is described by following expression:

$$\frac{1}{\sqrt{\zeta}} = 2.5 \lg \left(Re \sqrt{\zeta} \right) - 0.36. \tag{12}$$

The curve 2 (fig. 7) correspond to the formula (12), the curve 3 is Virk's asymptotic law [3] and the curve 1 is the logarithmic law of resistance for a smooth pipe [8], accordingly.

As shown from figure 7, experimental data of hydraulic resistance factor found on the basis of trial tests are good agreement with logarithmic dependence (12).

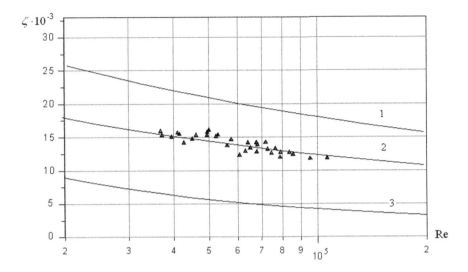

Fig. 7. Resistance factor vs Reynolds number

On the basis of the analysis of the pilot and calculated data it is found, that heat exchange between a oil-mixture flow and pipe wall can be defined by the formula:

$$Nu = \frac{Pr\, Re\sqrt{\zeta/2}}{4.24\ln\left(Re\sqrt{\zeta/16}\right) + 25\, Pr^{2/3} + 4.24\ln Pr - 20.2}. \tag{13}$$

Where numbers Re, Pr are determined by rheological properties of oil-mixture at presence of the anti-turbulent additives.

The received laws of resistance and heat exchange of pipe with additives allow leading heat-hydraulic calculation of the main oil pipelines.

2 Mathematical model of reforming process in the industrial reactor

Catalytic reforming is one of leading processes of the oil refining and petro-chemical industry. The basic purpose of catalytic reforming is reception of aromatic hydrocarbons and high octane gasolines. The process is carried out in not warmed system of consistently connected reactors with a motionless layer of the catalyst. There is product warming between reactors.

The known computational methods of reforming reactors are conducted by zero-dimensional scheme [13]- [14]. The zero-dimensional approaches calculate balance of material and heat flows for an input and output of the reactor, they does not allow to define distributions of flow velocity, concentration and temperature of a reaction mixture in a catalyst bed.

For optimum process control of reforming it is necessary to know laws of reaction mixture conversion in a catalyst bed. This operation can be executed by detailed calculation of mathematical model of aerodynamics, mass and heat transfer with account of catalytic chemical reactions.

The industrial reforming reactor with radial input of raw is shown in figure 8. Reactor represents a vessel with the internal perforated glass 3 where load the catalyst 2. Gas-raw mixture is loaded through input point 8 in annual gap between brick-lining 10 and glass 3, passes on radial direction through the catalyst layer and is offloaded through the perforated pipe 7.

The composition of a petrol fraction is shown in table 4. The mix contains a circulating gas with hydrogen for reduction of coking deposits in the catalyst [13]-[14].

Table 4. The composition of petrol fraction of catalytic reforming

Fractional structure, %						Hydrocarbon structure, %		
ρ_{277}^{293}	B.b.	10%	50%	90%	F.b.	Aromatics	Naphthenes	Paraffins
0.7288	329	348	385	428	453	12	38	50

The reacting mix cooperates with a surface of the catalyst, where complex chemical transformations proceed. According to Smith's kinetic model naphthen components of oil fraction turn to aromatics and paraffins, paraffin components to naphthenes, aromatic components to naphthenes, and also hydrocraking of naphthen and paraffin hydrocarbons occur.

The chemical reactions proceeding in a catalyst layer [15]:

conversion of naphthenes to aromatics:

$$C_n H_{2n} \leftrightarrow C_n H_{2n-6} + 3H_2, \tag{14}$$

conversion of naphthenes to paraffins:

$$C_n H_{2n} + H_2 \leftrightarrow C_n H_{2n+2}, \tag{15}$$

hydrocracking of naphthenes:

$$C_n H_{2n} + \frac{n}{3} H_2 \rightarrow \frac{n}{15} \left(CH_4 + C_2 H_6 + C_3 H_8 + C_4 H_{10} + C_5 H_{12} \right), \tag{16}$$

hydrocracking of paraffins:

$$C_n H_{2n+2} + \frac{n-3}{3} H_2 \rightarrow \frac{n}{15} \left(CH_4 + C_2 H_6 + C_3 H_8 + C_4 H_{10} + C_5 H_{12} \right). \tag{17}$$

Some of the reactions proceed with heat release, and the some of it's with heat absorption.

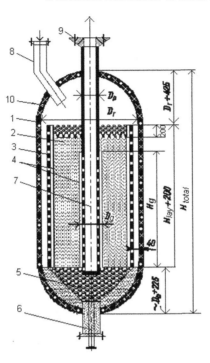

Fig. 8. Scheme of the reactor with radial load of raw material: 1 body, 2-catalyst, 3-perforated glass with a grid, 4-grid, 5-porcelain balls, 6-gas ejection point, 7-perforated pipe, 8-point for raw material load, 9-point for an output of reaction product, 10 – lining

The stream of a reacting mix proceeds on a ring collector and through the perforated lateral wall and flows into a layer of the catalyst. There is a distribution of a reacting mix on the ring collector during movement process. The law of stream distribution is beforehand unknown and depends on aerodynamics of flow in collectors, resistances of the catalyst layer, the perforated lateral wall and the grid supporting the catalyst layer.

Reforming process is considered in cylindrical coordinates system. The OZ axis goes on axes of a reactor, and the OR axis - in a radial direction.

The mathematical model is constructed in the following assumptions: 1) flow is laminar, gas is viscous; 2) process proceeds symmetrically concerning a longitudinal axis of a reactor; 3) mix flow in the catalyst layer satisfies to the nonlinear law of a filtration.

The mathematical model consists of equations of movement and heat-mass transfer with chemical reactions:

$$\rho u \frac{\partial u}{\partial z} + \rho v \frac{\partial u}{\partial r} = -\frac{\partial p}{\partial z} + 2\frac{\partial}{\partial z}\left(\mu \frac{\partial u}{\partial z}\right) - \frac{2}{3}\frac{\partial}{\partial z}\left(\mu \, div\, \overrightarrow{V}\right) +$$
$$+\frac{1}{r}\frac{\partial}{\partial r}\left[\mu r \left(\frac{\partial u}{\partial r} + \frac{\partial v}{\partial z}\right)\right] - \left(\zeta_1 + \zeta_2|\mathbf{V}|\right)\rho u, \tag{18}$$

$$\rho u \frac{\partial v}{\partial z} + \rho v \frac{\partial v}{\partial r} = -\frac{\partial p}{\partial r} + \frac{2}{r}\frac{\partial}{\partial r}\left(\mu r \frac{\partial v}{\partial r}\right) - \frac{2}{3}\frac{\partial}{\partial r}\left(\mu div \vec{V}\right) +$$
$$+\frac{\partial}{\partial z}\left[\mu\left(\frac{\partial u}{\partial r} + \frac{\partial v}{\partial z}\right)\right] - \frac{2v}{r^2}\mu - (\zeta_1 + \zeta_2|\mathbf{V}|)\,\rho v, \tag{19}$$

$$\frac{\partial \varepsilon \rho u}{\partial z} + \frac{1}{r}\frac{\partial(\varepsilon \rho v r)}{\partial r} = 0, \tag{20}$$

$$\rho u \frac{\partial c_N}{\partial z} + \rho v \frac{\partial c_N}{\partial r} = \frac{\partial}{\partial z}\left(\rho D_z \frac{\partial c_N}{\partial z}\right) + \frac{1}{r}\frac{\partial}{\partial r}\left(\rho D_r r \frac{\partial c_N}{\partial r}\right) + W_N, \tag{21}$$

$$\rho u \frac{\partial c_A}{\partial z} + \rho v \frac{\partial c_A}{\partial r} = \frac{\partial}{\partial z}\left(\rho D_z \frac{\partial c_A}{\partial z}\right) + \frac{1}{r}\frac{\partial}{\partial r}\left(\rho D_r r \frac{\partial c_A}{\partial r}\right) + W_A, \tag{22}$$

$$\rho u \frac{\partial c_P}{\partial z} + \rho v \frac{\partial c_P}{\partial r} = \frac{\partial}{\partial z}\left(\rho D_z \frac{\partial c_P}{\partial z}\right) + \frac{1}{r}\frac{\partial}{\partial r}\left(\rho D_r r \frac{\partial c_P}{\partial r}\right) + W_P, \tag{23}$$

$$\rho u c\frac{\partial T}{\partial z} + \rho v \frac{\partial T}{\partial r} = \frac{\partial}{\partial z}\left(\lambda_z \frac{\partial T}{\partial z}\right) + \frac{1}{r}\frac{\partial}{\partial r}\left(\lambda_r r \frac{\partial T}{\partial r}\right) + I_{N\to A}\left(\omega_1 - \omega_5\right) +$$
$$+I_{P\to N}\left(\omega_2 - \omega_4\right) + I_{N\to G}\omega_3 + I_{P\to G}\omega_6. \tag{24}$$

The system of equations (18) - (24) is solved at following boundary conditions:
on entrance section of ring collector $\Gamma 1$:

$$u = u(r), \quad v = 0, \quad c_N = c_N^0, \quad c_A = c_A^0, \quad c_P = c_P^0, \quad T = T_0, \tag{25}$$

on firm walls of collector $\Gamma 2$:

$$u = 0,\ v = 0,\quad \frac{\partial c_N}{\partial r} = 0,\quad \frac{\partial c_A}{\partial r} = 0,\quad \frac{\partial c_P}{\partial r} = 0,\quad \frac{\partial T}{\partial r} = 0, \tag{26}$$

on symmetry axis of reactor $\Gamma 3$:

$$\frac{\partial u}{\partial r} = 0,\ v = 0,\quad \frac{\partial c_N}{\partial r} = 0,\quad \frac{\partial c_A}{\partial r} = 0,\quad \frac{\partial c_P}{\partial r} = 0,\quad \frac{\partial T}{\partial r} = 0, \tag{27}$$

on exit section of internal collector $\Gamma 4$:

$$\frac{\partial u}{\partial z} = 0,\ v = 0,\quad \frac{\partial c_N}{\partial z} = 0,\quad \frac{\partial c_A}{\partial z} = 0,\quad \frac{\partial c_P}{\partial z} = 0,\quad \frac{\partial T}{\partial z} = 0, \tag{28}$$

where z, r – cylindrical axes, m, u, v – longitudinal and transversal components of velocity, m/s, p – pressure, Pa, μ – coefficient of dynamic viscosity, kg / (m·s), ρ – density of raw, kg/m^3, c_N, c_A, c_P – concentrations of naphthenes, aromatics, paraffins accordingly, kmol/kg, c – thermal capacity of gas, kJ / (kg·K), T – temperature, K, D – effective diffusion coefficient, m^2/s, λ – effective thermal conductivity, Wt/ (m·K), I – heat of reaction, kJ/kmol.

At $\varepsilon = 1$ the system of equations transfers to the usual Navier-Stocks equations and it applicable to flow in free parts of reactor, and at $\varepsilon < 1$ it describes flow in the catalyst layer [16]-[17].

The system of equations (18) - (24) is solved by the numerical method in variables of stream function and vortex strength.

The distribution of a streamline is shown in Fig. 9a at G=3600ton/day. The streamlines illustrate a flow pattern of reacting mix in the ring collector, a catalyst bed and on target collector. The distribution law of a reacting mix in the catalyst bed depends on flow resistance in the input ring collector, in the catalyst bed and on target collector. I.e. the distribution of streamlines is determined depending on values of the indicated resistance. Streamlines in an initial site of a distributing collector flow not orthogonally into a granular layer of the catalyst, i.e. the impulse of longitudinal flow is big also causes longitudinal transfer of movement quantity on an entrance zone of catalyst. The longitudinal impulse of a stream is reduced in process of movement on a ring collector and causes reduction of a longitudinal overflow. In a final part of a distributing collector the streamlines flow orthogonally into a layer

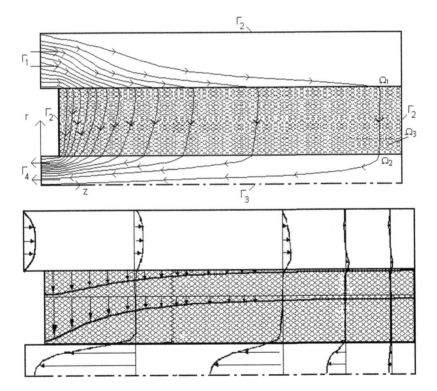

Fig. 9. Distribution of stream line (a) and distribution of speed (b) in an active zone of a reactor, productivity of reactor G=3600 ton/day

of the catalyst. The reacting mix flows practically orthogonally in a radial direction, however longitudinal overflows in a flow direction of a collecting collector are observed on an output of a layer, i.e. the impulse of longitudinal flow in a collecting collector affects. The streamlines in a layer of the catalyst are located more densely in a longitudinal direction in 1/3 parts of active zone. It means, that speed of a filtration of a reacting mix in this part more than on other parts and takes place non-uniform distribution of filtration rate on height of catalyst layer.

Figure 9 b shows the distribution of filtration speed. In an entrance part of a layer non-uniformity of radial speed in a longitudinal direction is insignificant. However in process of movement on thickness of the layer non-uniformity of filtration speed becomes significant: at first, because of longitudinal overflows of a mix, secondly, because of reduction of radial coordinate. It is easy to notice, that radial speed changes in inverse proportion to radius. From figure 9 b it is well visible, how there is a distribution of a reacting mix in a ring collector, gradual decrease and deformation of longitudinal speed diagram with occurrence of a zone of returnable flow. Also it is well visible how there is a gathering a reacting mix in a collecting collector both gradual increase and deformation of longitudinal speed. All the calculated data evidently show aerodynamics of flow in collectors and in a layer of the catalyst, describe laws of movement of a reacting mix in a reforming reactor.

Distributions of concentration of components on length of the catalyst layer are shown in figures 10-12. Distributions of concentration on medium and exit layers of the catalyst are constructed. The calculate data of concentration of the components according to Smith's model define transformations of naphthen hydrocarbons to aromatic.

There are transformations naphthenic and aromatic hydrocarbons inside a layer during the process of movement of a reacting mix. And distributions of concentration of components on length catalyst layer are non-uniform. It speaks that filtration speed is heterogeneity. Great values of filtration speed

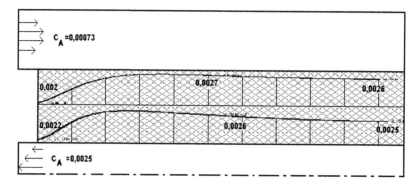

Fig. 10. Distribution of aromatic hydrocarbons on length of catalyst layer of reactor, value of concentration are given in kmol/kg

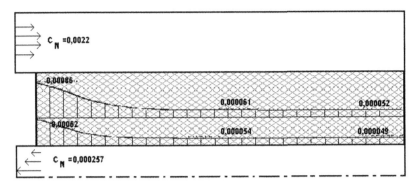

Fig. 11. Distribution of naphthenic hydrocarbons on length of catalyst layer of reactor, value of concentration are given in kmol/kg

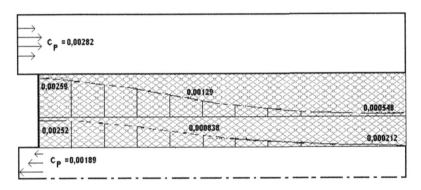

Fig. 12. Distribution of paraffinic hydrocarbons on length of catalyst layer of the first reactor, value of concentration are given in kmol/kg

cause increase in mass transfer rate between a stream and catalytic layer in comparison with chemical reactions rate in grain of the catalyst. There is the reacting mix has not time to react in the field of the big filtration speed because of small time of contact, and there is more intensive transformation of hydrocarbons in the field of small speed of the stream.

The distribution of concentration of paraffin hydrocarbons on length of catalytic layer in different sections on radius of a reactor is shown in Fig. 12. It is easy to notice, that non-uniformity in distribution of concentration is caused by heterogeneity of distribution of filtration speed. As show calculations, as a result of chemical transformations in catalyst layer the increase in aromatic hydrocarbons, reduction of naphthen and paraffin hydrocarbons are observed. That is coordinated to the experimental data

In the conclusion it is possible to ascertain about creation of full-scale mathematical model of reforming process in industrial catalytic reactors. Results of research allow to carry out the detailed analysis of reforming process according to production schedules of industrial reactors and to find optimum operating modes.

References

1. Toonder JMJ, Hulsen MA, Kuiken GDC, Nieuwstadt FTM (1997) J Fluid Mech 337:193–231
2. Yu B, Kawaguchi Y (2003) DNS of turbulent heat transfer of drag-reducing flow with surfactant additives. In: Proc. of 4th Int. Symp. on turbulence, heat and mass transfer. Antalya, Turkey
3. Virk PS (1975) AIChE Journal 21:625–653
4. Prohorov AD, Chelintsev SN, etc (2000) Transport i khranenie nefti 3:8–9 (in Russian)
5. Belousov YuP (1986) Anti-turbulent additives for hydrocarbonic fluids. Nauka, Moscow (in Russian)
6. Makarov SP, Fokin SM, Eroshkina II, et al (2000) Transport i khranenie nefti 4:14–17 (in Russian)
7. Loitsyanski LG (1978) Mechanics of fluid anf gas. Nauka, Moscow (in Russian)
8. Shlichting G (1974) Theory of a boundary layer. Nauka, Moscow (in Russian)
9. Gosman AD, Pun WM, Runchal AK, Spalding DB, Wolfshtein M (1969) Heat and mass transfer in recirculating flows. Academic Press, London, New York
10. Chen HC, Patel VC (1988) AIAA J 26:641–648
11. Chien KY (1982) AIAA J 20:33–38
12. Myoung HK, Kasagi N (2003) JSME Int J 33:63–72
13. Kuznetsov AA, Kagermanov SM, Sudakov EN (1974) Calculations of processes and devices of petroleum-refining industry. Chemistry, Leningrad (in Russian)
14. Skoblo AI, Tregubova IA, Molokanov JuK (1982) Process and devices of oil refining and petrochemical industry. Chemistry, Moscow (in Russian)
15. Smith RB (1959) Chem Eng Prog 55: 76-80.
16. Ershyn SA, Zhapbasbaev UK, Balakaeva GT (1997) Theory and calculation of catalytic clearing devices. KazNU, Almaty (in Russian)
17. Aerov ME, Todes OM, Narynski DA (1979) Device with stationary granular layer. Chemistry, Leningrad (in Russian)

Large-eddy simulations for tundish and airfoil flows

N. A. Alkishriwi, Q. Zhang, M. Meinke, and W. Schröder

Institute of Aerodynamics, RWTH Aachen University, Wüllnerstr. zw. 5 u. 7, 52062 Aachen, Germany office@aia.rwth-aachen.de

Summary. Large-eddy simulations (LES) of two complex flow problems, a continuous tundish flow and the flow around multi-element airfoils are presented. The numerical computations are performed by solving the conservation equations for compressible fluids. An implicit dual time stepping scheme combined with low Mach number preconditioning and a multigrid accelerating technique is developed for LES computations. The method is validated by comparing data of turbulent pipe flow at Re_τ=1280 and cylinder flow at Re=3900 at different Mach numbers with experimental findings from the literature. Finally, the characteristics of the flow in a one-strand tundish is analyzed and a solution for a flow around a two-element airfoil as well as a zonal solution for the trailing edge region are discussed.

1 Introduction

In many engineering problems compressible and nearly incompressible flow regimes occur simultaneously. For example low speed flows, which may be compressible due to surface heat transfer or volumetric heat addition. The numerical analysis of such flows requires to solve the viscous conservation equations for compressible fluids to capture the essential effects.

When a compressible flow solver is applied to a nearly incompressible flow, its performance can deteriorate in terms of both speed and accuracy [1]. It is well known that most compressible codes do not converge to an acceptable solution when the Mach number of the flow field is smaller than $\mathcal{O}(10^{-1})$. The main difficulty with such low speed flows arises from the large disparity between the wave speeds. The acoustic wave speed is $|u \pm c|$, while entropy or vorticity waves travel at $|u|$, which is quite small compared to $|u \pm c|$. In explicit time-marching codes, the acoustic waves define the maximum time step, while the convective waves determine the total number of iterations such that the overall computational time becomes large for small Mach numbers.

Different methods have been proposed to solve such mixed flow problems by modifying the existing compressible flow solvers. One of the most popular approaches is to use low Mach number preconditioning methods for compressible codes [1]. The basic idea of this approach is to modify the time marching behavior of the system of equations without altering the steady state solution. This is, however, only useful when the steady state solution is sought. A straightforward extension of the preconditioning approach to unsteady flow problems is achieved when it is combined with a dual time stepping technique.

This idea is followed in the present paper, i.e., a highly efficient large-eddy simulation method is described based on an implicit dual time stepping scheme combined with preconditioning and multigrid. This method is validated by well-known case studies [2] and finally, the preconditioned large-eddy simulation method is used to investigate the flow field in a continuous casting tundish and around airfoils in high lift configuration.

In continuous casting of steel, the tundish enables to remove nonmetallic inclusions from molten steel and to regulate the flow from individual ladles to the mold. There is two types of flow conditions in a tundish: steady-state casting where the mass flow rate through the shroud m_{sh} is equal to the mass flow rate through the submerged entry nozzle into the mold m_{SEN}, and transient casting in which the mass of steel in a tundish varies in time during the filling or draining stages. The motion of the liquid steel is generated by jets into the tundish and continuously casting mold. The flow regime is mostly turbulent, but some turbulence attenuation can occur far from the inlet. The characteristics of the flow in a tundish include jet spreading, jet impingement on the wall, wall jets, and an important decrease of turbulence intensity in the core region of the tundish far from the jet [3, 4].

In previous studies, a large amount of research has been carried out to understand the physics of the flow in a tundish mainly through numerical simulations based on the Reynolds-averaged NAVIER–STOKES (RANS) equations plus an appropriate turbulence model. To gain more knowledge about the transient turbulence process, which cannot be achieved via RANS solutions, large-eddy simulations of the tundish flow field are performed.

Turbulent flow near the trailing edge of a lifting surface generates intense, broadband scattering noise as well as surface pressure fluctuations. The accuracy of the trailing-edge noise prediction depends on the prediction method of the noise-generating eddies over a wide range of length scales. Recent studies indicate that promising results can be obtained when the unsteady turbulent flow fields are computed via large eddy simulation (LES). The LES data can be used to determine acoustic source functions that occur in the acoustic wave propagation equations which have to be solved to predict the aero-acoustical field [5].

The turbulence embedded within a boundary layer is known to radiate quadrupole noise which in turn is scattered at the trailing edge [6]. The latter gives rise to an intense noise radiation which is called trailing-edge noise. Turbulence is an insufficient radiator of sound, particularly at low Mach number, $M_\infty \leq 0.3$, meaning that only a relatively small amount of energy is radiated away as sound from the region of the turbulent flow. However, if the turbulence interacts with a trailing edge in the flow, scattering occurs which changes the inefficient quadrupole radiation into a much more efficient dipole radiation [6].

The noise generated by an airfoil is attributed to the instability of the upper and lower surface boundary layers and their interactions with the trailing edge [7]. The edge is usually a source of high-frequency sound associated with smaller-scale components of the boundary layer turbulence. Low-frequency contributions from a trailing edge, that may in practice be related to large-scale vortical structures shed from an upstream perturbation, are small because the upwash velocity they produce in the neighborhood of the edge tends to be canceled by that produced by vorticity shed from the edge [7].

In order to predict the trailing-edge aeroacoustics, one should use a hybrid two-step approach that connects the large-eddy simulation and the aeroacoustic theories [5]. For this purpose, a large-eddy simulation is carried out to simulate the three-dimensional compressible turbulent boundary layer past an airfoil-flap configuration and and airfoil trailing edge. The present results will comprise an analysis of the turbulent scales upstream of the trailing-edge and the trailing-edge eddies generated in the near-field wake based on LES findings. The simulation provides the data that allows the acoustic source functions of the acoustic wave equations to be evaluated.

After a concise presentation of the governing equations the implementation of the preconditioning in the LES context using the dual time stepping technique is described. Then, the discretization and the time marching solution technique within the dual time stepping approach are discussed. Then numerical results of the validation cases, the tundish and the airfoil simulations are presented.

2 Governing equations

The governing equations are the unsteady three-dimensional compressible NAVIER-STOKES equations written in generalized coordinates $\xi_i, i = 1, 2, 3$

$$\frac{\partial \mathbf{Q}}{\partial t} + \frac{\partial(\mathbf{F}_{c_i} - \mathbf{F}_{v_i})}{\partial \xi_i} = 0, \qquad (1)$$

where the quantity $\mathbf{Q}$ represents the vector of the conservative variables and $\mathbf{F}_{c_i}$, $\mathbf{F}_{v_i}$ are inviscid and viscous flux vectors, respectively. As mentioned before, preconditioning is required to provide an efficient and accurate method of solution of the steady NAVIER–STOKES equations for compressible flow at low Mach numbers. Moreover, when unsteady flows are considered, a dual time stepping technique for time accurate solutions is used. In this approach, the solution at the next physical time step is determined as a steady state problem to which preconditioning, local time stepping and multigrid are applied.

Introducing of a pseudo-time τ in (1), the unsteady two-dimensional governing equations with preconditioning read

$$\mathbf{\Gamma}^{-1}\frac{\partial \mathbf{Q}}{\partial \tau} + \frac{\partial \mathbf{Q}}{\partial t} + \mathbf{R} = 0, \qquad (2)$$

where $\mathbf{R}$ represents

$$\mathbf{R} = \left(\frac{\partial \mathbf{E}_c}{\partial \xi} + \frac{\partial \mathbf{F}_c}{\partial \eta} + \frac{\partial \mathbf{E}_v}{\partial \xi} + \frac{\partial \mathbf{F}_v}{\partial \eta} \right) \qquad (3)$$

and $\mathbf{\Gamma}^{-1}$ is the preconditioning matrix, which is to be defined such that the new eigenvalues of the preconditioned system of equations are of similar magnitude. In this study, a preconditioning technique from Turkel [1] has been implemented.

It is clear that only the pseudo-time terms in (2) are altered by the preconditioning, while the physical time and space derivatives retain their original form. Convergence of the pseudo-time within each physical time step is necessary for accurate unsteady solutions. This means, the acceleration techniques such as local time stepping and multigrid can be immediately utilized to speed up the convergence within each physical time step to obtain an accurate solution for unsteady flows. The derivatives with respect to the physical time t are discretized using a three-point backward difference scheme that results in an implicit scheme, which is second-order accurate in time

$$\frac{\partial \mathbf{Q}}{\partial \tau} = \mathbf{RHS} \qquad (4)$$

with the right-hand side

$$\mathbf{RHS} = -\mathbf{\Gamma} \left(\frac{3\mathbf{Q}^{n+1} - 4\mathbf{Q}^n + \mathbf{Q}^{n-1}}{2\Delta t} + \mathbf{R}(\mathbf{Q}^{n+1}) \right).$$

Note that at $\tau \to \infty$ the first term on left-hand side of (2) vanishes such that (1) is recovered. To advance the solution of the inner pseudo-time iteration, a 5-stage Runge-Kutta method in conjunction with local time stepping and multigrid is used. For stability reasons the term $\frac{3\mathbf{Q}^{n+1}}{2\Delta t}$ is treated implicitly within the Runge-Kutta stages yielding the following formulation for the lth stage

$$\mathbf{Q}^0 = \mathbf{Q}^n$$
$$\vdots$$

$$\mathbf{Q}^l = \mathbf{Q}^0 + \alpha_l \Delta \tau \left[\mathbf{I} + \frac{3\Delta \tau}{2\Delta t} \alpha_l \mathbf{\Gamma} \right]^{-1} \mathbf{RHS} \qquad (5)$$

$$\vdots$$
$$\mathbf{Q}^{n+1} = \mathbf{Q}^5.$$

The additional term means that in smooth flows the development in pseudo-time is proportional to the evolution in t.

3 Numerical method

The governing equations are the NAVIER–STOKES equations filtered by a low-pass filter of width Δ, which corresponds to the local average in each cell volume. The monotone integrated large-eddy simulations (MILES) approach is used to implicitly model the small scale motions through the numerical scheme.

The approximation of the convective terms of the conservation equations is based on a modified second-order accurate AUSM scheme using a centered 5-point low dissipation stencil [8] to compute the pressure derivative in the convective fluxes. The pressure term contains an additional expression, which is scaled by a weighting parameter χ that represents the rate of change of the pressure ratio with respect to the local Mach number. This parameter determines the amount of numerical dissipation to be added to avoid oscillations that could lead to unstable solutions. The parameter χ was chosen in the range $0 \leq \chi \leq \frac{1}{400}$. The viscous stresses are discretized to second-order accuracy using central differences, i.e., the overall spatial approximation is second-order accurate.

A dual time stepping technique is used for the temporal integration. In this approach, the solution at the next physical time step is determined as a steady state problem to which preconditioning, local time stepping and multigrid are applied. A 5-stage Runge-Kutta method is used to propagate the solution from time level n to n+1. The Runge-Kutta coefficients $\alpha_l = (\frac{6}{24}, \frac{4}{24}, \frac{9}{24}, \frac{12}{24}, \frac{24}{24})$ are optimized for maximum stability of a centrally discretized scheme. The physical time derivative is discretized by a backward difference formula of second-order accuracy. The method is formulated for multi-block structured curvilinear grids and implemented on vector and parallel computers.

3.1 Validation

To validate the efficiency and the accuracy of the method, large-eddy simulations of turbulent pipe flow at a Reynolds number $Re_\tau = 1280$ based on the friction velocity u_τ, which corresponds to a diameter D based Reynolds number $Re_D = u_{cl}D/\nu = 22550$, are investigated. The Mach number based on the centerline velocity of the pipe is set to $Ma = 0.02$ and the physical time step $\Delta t = 0.01$. The comparison of the pure explicit LES results from [9, 10] and the LES findings of the implicit method in Fig. 1 show good agreement for the mean velocity profiles and the turbulence intensity distributions.

The flow around a cylinder at a diameter based $Re_D = 3900$ is performed at a freestream Mach number $M_\infty = 0.05$ and a physical time step $\Delta t = 0.02$. Fig. 2 shows the streamwise velocity distribution on the centerline in the wake of the cylinder compared with the LES distribution of a pure explicit scheme without preconditioning at a Mach number $M_\infty = 0.1$ and with experimental data from the literature. The profiles of the velocity fluctuations of the streamwise and vertical components at $X/D = 1.54$ as a function of Y/D

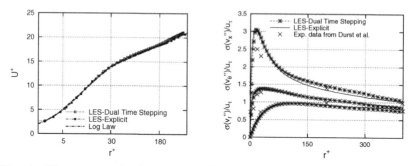

Fig. 1. Time averaged velocity profile (left) and turbulence intensities (right) of a turbulent pipe flow at Re_τ=1280

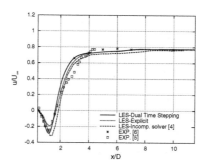

Fig. 2. Streamwise velocity as a function of X/D on the centerline $Y/D = 0$, $Z/D = 0$ in the cylinder wake at $Re_D = 3900$

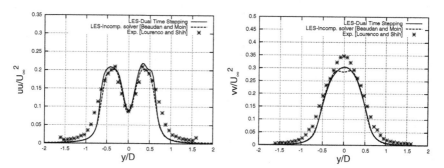

Fig. 3. Streamwise (left) and vertical (right) velocity fluctuations as a function of Y/D in the cylinder wake at $X/D = 1.54$

presented in Fig. 3 corroborate the correspondence with the measurements cited in [11].

The efficiency analysis is performed for different parameters such as the size of the physical time step, the required residual constraint for the inner pseudo-time iteration, the Mach number and the Reynolds number. Figs. 4 and 5 demonstrate the impact of the size of the physical time step and the

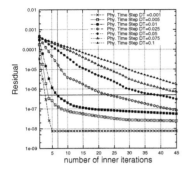
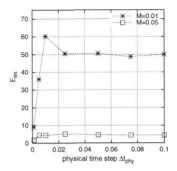

Fig. 4. Convergence of the inner iterations at $Ma = 0.01$ for channel flow computations at $Re_\tau = 590$

Fig. 5. Efficiency E_{im} as a function of Δt_{phy} at $Ma = 0.01$ and $Ma = 0.05$ for channel flow computations at $Re_\tau = 590$

Mach number on the efficiency of the implicit scheme. It is evident that as the size of the physical time step increases, the required number of iterations in the inner pseudo-time cycle grows. Fig. 5 demonstrates the impact of the Mach number on the efficiency of the implicit scheme. The results, which are based on the simulations of turbulent channel flow at $Re_\tau = 590$ and $Ma = 0.01$, show a speedup in the range of 9 to 60 compared to the explicit scheme for a reduction of two orders of magnitude. Note, that for a reduction of three orders of magnitude this speedup is lowered to a range of 6 to 26.

4 Tundish flow

In the following the numerical setup of the tundish flow is briefly described. The geometry and the flow configuration for the simulation are shown in Fig. 6.

The numerical simulations consist of two simultaneously performed computations on the one hand, the pipe flow is calculated to provide time-dependent inflow data for the jet into the tundish and on the other hand, the

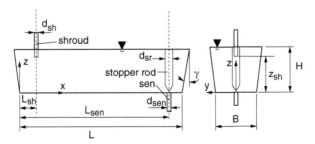

Fig. 6. Sketch of the tundish geometry with the main parameters

Fig. 7. Computational mesh for the tundish flow. 12 million grid points in 28 blocks

flow field within the tundish. The geometrical values and the flow parameters of the tundish are given in Table 1. The computational domain is discretized by 12 million grid points, Fig. 7, 5 million of which are located in the jet domain to resolve the essential turbulent structures. Since the jet possesses the major impact on the flow characteristics in the tundish, it is a must to determine in great detail the interaction between the jet and the tundish flow.

Table 1. Physical parameters of the tundish flow

tundish length	L	1.847 m
tundish width	B	0.459 m
tundish height	H	0.471 m
inclination of side walls	γ	7°
diameter of the shroud	d_{sh}	0.04 m
height between bottom shroud	z_{sh}	0.352 m
diameter of the SEN	d_{sen}	0.041 m
diameter of the stopper rod	d_{sr}	0.0747 m
hydraulic diameter	d_{hyd}	0.6911 m
Reynoldsnumber based on the jet diameter d_{sh}	Re	25000

The boundary conditions consist of no-slip conditions on solid walls. At the free surface, the normal velocity components and the normal derivatives of all remaining other variables are set zero. An LES of the impinging jet requires a prescription of the instantaneous flow variables at the inlet section of the jet. To determine those values a slicing technique based on a simultaneously conducted LES of a fully developed turbulent pipe flow is used.

The typical structure of the flow field in the tundish is shown in Figs. 8-11. These figures show the computed flow pattern in different locations within the tundish. Figs. 8 and 9 visualize the jet flow into the tundish. It is clear that

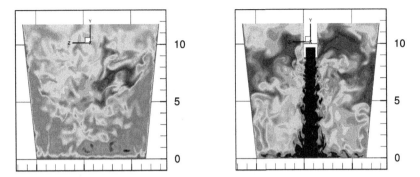

Fig. 8. Instantaneous entropy contours at $X/L_1=0$ (left) and in the center plane of the jet (right)

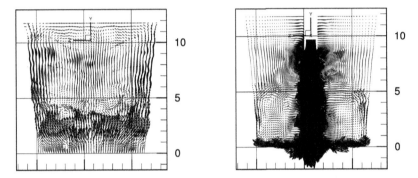

Fig. 9. Instantaneous velocity vectors at $X/L_1=0$ (left) and in the center plane of the jet (right)

the turbulent flow contains a wide range of length scales. Large eddies with a size comparable to the diameter of the pipe occur together with eddies of very small size. The figures evidence the jet spreading, jet impingement on the wall, and wall jets. The jet ejected from the ladle reaches the bottom of the tundish at high velocities, spreads in all directions and then mainly flows along the side walls of the tundish. Such a flow pattern leads to nonmetallic inclusions, which should be avoided.

Figs. 10 and 11 represent the flow field with the help of λ_2 contours and planes with the magnitude of the velocity for a time averaged and an instantaneous solution. These figures illustrate the vortex dominated flow in the tundish. It can be seen that there are two strong vortices at the side walls of the inlet region of the tundish, visible in the time averaged flow field. The instantaneous flow field shows a fully turbulent flow in the inlet region, where the turbulence decays when the flow passes half through the tundish.

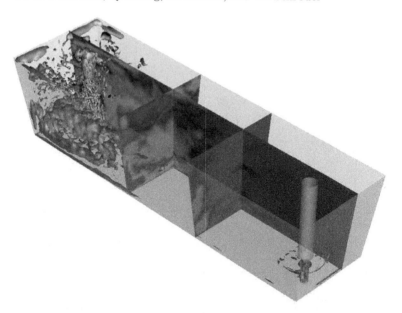

Fig. 10. λ_2 contours of the time averaged flow field in the tundish, color coded with the magnitude of the velocity

Fig. 11. λ_2 contours in the tundish, color coded with the magnitude of the velocity

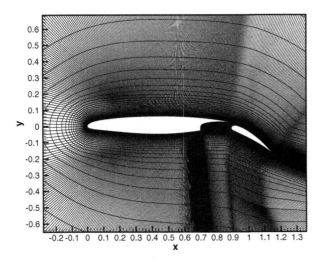

Fig. 12. Computational mesh in the x-y plane, which shows every second grid point

5 Airfoil flow

One of the primary conclusions from the European LESFOIL project is that an adequate numerical resolution especially in the near wall region is required for a successful LES. On meshes which do not resolve the viscous near-wall effects, neither SGS models nor wall models were able to remedy these deficiencies [12]. According to the experience from LES of wall bounded flows, the resolution requirements of a wall resolved LES are in the range of $\Delta x^+ \approx 100, \Delta y^+ \approx 2$ und $\Delta z^+ \approx 20$, where x, y, z denote the streamwise, normal, and spanwise coordinates, respectively. During the mesh generation process, it became clear that it would lead to meshes with unmanageable total grid point numbers, if these requirements are to be followed strictly. For the preliminary study, the streamwise resolution is approximately $\Delta x^+ \approx 200 \sim 300$, whereas the resolution is set to $\Delta y^+_{min} \approx 2$ and $\Delta z^+ \approx 20 \sim 25$ in the wall normal and in the spanwise direction, respectively. The mesh of the preliminary study is shown in Fig. 12.

Table 2. Computational domain and grid point distribution, SWING+ Airfoil $(Re = 10^6)$

L_X	L_Y	L_Z	$N_x \times N_y$	N_z	Total Grid Points
-4.0 ... 5.0	-4.0 ... 4.0	0 ... 0.0032	241.320	17	4.102.440

The extent of the computational domain is listed in Tab. 2. The spanwise extent of the airfoil is 0.32 percent of the chord length. A periodic boundary condition is used for this direction. The relatively small spanwise extent is chosen because of the following two reasons. On the one hand, it was shown for flat plate flows that due to the high Reynolds number of $Re = 1.0 \times 10^6$, two-point correlations decay to zero already in about 250 wall units [13] and on the other hand, since a highly resolved mesh was used in the wall normal and tangential direction the computational domain must be limited to a reasonable size to reduce the overall computational effort. Towards the separation regions in the flap cove and at the flap trailing edge, the boundary layer thicknesses and the characteristic sizes of turbulent structures will increase. Therefore, especially for these areas, the spanwise extent of the computational domain needs to be increased in further computations. A no-slip boundary condition is applied at the airfoil surface, and a non-reflecting boundary condition according to Poinsot & Lele [14] is used for the far field.

Another possibility to enlarge the spanwise extent without increasing the total number of grid points and as such the overall computational effort, is to focus the analysis on the zone of the flow field that is of major interest for the sound propagation. For the airfoil flow problem this means that a local analysis of the flow in the vicinity of the trailing edge has to be pursued. To do so, we apply the rescaling method by El-Askary [13] which is valid not only in compressible flows but also in flows with weak pressure gradients. This procedure is a consistent continuation of the approach that was already successfully applied in the analysis of the trailing edge flow of a sharp flat plate [5, 15, 16, 17, 18].

In the following, we discuss first the LES of the airfoil-flap configuration and then turn to the zonal analysis of the airfoil trailing edge flow. In both problems numerical and experimental data are juxtaposed.

In the flow over the SWING+ airfoil with deflected flap, separations exist in the flap cove and at the flap trailing edge. In the current numerical simulation, both separation regions are well resolved. They are visible in the time and spanwise averaged streamlines of the numerical simulation (Fig. 13).

The turbulent flow is considered to be fully developed such that the numerical flow data can be time averaged over a time period of $\Delta T = 3.0\ c/u_\infty$, and then spanwise averaged to obtain mean values. The distribution of the mean pressure coefficient c_p over the airfoil surface is plotted in Fig. 14. A good agreement with the experimental data [19] is achieved. It is worth mentioning, that the onset of the turbulent flow separation and the size of the separation bubble in the vicinity of the flap trailing edge are captured quite exactly by the numerical simulation. This can be seen by the plateau in the c_p-distribution near the trailing edge of the flap.

In the experiments carried out by the Institute of Aerodynamics and Gasdynamics of the University Stuttgart, profiles of the wall tangential velocity were measured at five locations two of which were on the airfoil and three on the flap [19]. The mean velocity profiles from the experiments and the

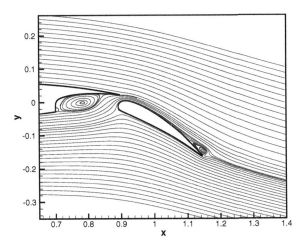

Fig. 13. Time and spanwise averaged streamlines

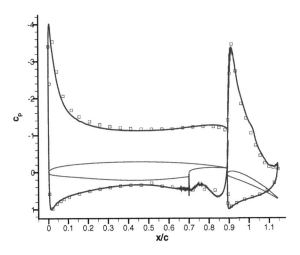

Fig. 14. Time and spanwise averaged pressure coefficients. Solid line: numerical data, Symbols: experimental data

numerical simulation are compared in Fig. 16. The qualitative trends of the velocity profiles agree with each other. Note that in the near wall region of position D, the measuring location lies in the separation area, where the hot

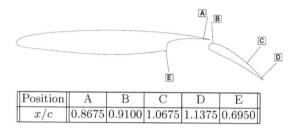

Position	A	B	C	D	E
x/c	0.8675	0.9100	1.0675	1.1375	0.6950

Fig. 15. Locations of the velocity measurements on the airfoil and flap

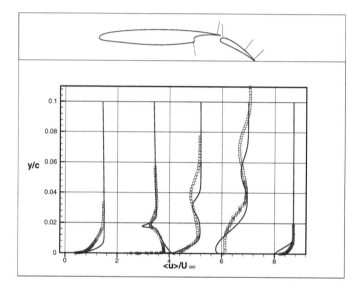

Fig. 16. Time and spanwise averaged profiles of the wall tangential velocity components. Solid line: numerical data, Symbols: experimental data

wire probe cannot obtain the correct velocity information. This is the reason for the different signs of the velocity profiles in the near wall region.

As can be expected from the good agreement of the c_p distribution over the airfoil, the lift coefficient agrees well with the experimental value. The drag coefficient, however, is over-predicted by the numerical simulation, which is not surprising due to the discrepancies in the velocity profiles.

We now turn to the discussion of the zonal large eddy simulation of an airfoil trailing-edge flow at an angle of attack of $\alpha = 3.3$ deg. The trailing-edge length constitutes 30% of the chord length c. The computational domain is shown in Fig. 17. All flow parameters are given in Tab. 3, in which Re_c is the Reynolds number based on the chord length. In the present simulation the inflow section is located at a position of an equilibrium turbulent boundary layer with a weak adverse pressure gradient. Therefore, all inflow data for the

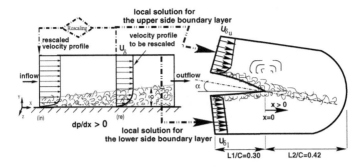

Fig. 17. Computational domain for the airfoil trailing edge flow (right) and inflow distribution from a flat plate boundary layer with an equilibrium adverse pressure gradient via the slicing technique

upper side boundary layer of the airfoil δ_u ($\delta_u = 0.01972c$) and for the lower side boundary layer δ_l ($\delta_l = 0.0094745c$) are extracted from two separate LES of flat plate boundary layers with an equilibrium pressure gradient, which use the new rescaling formulation for a variable pressure pressure gradient. A total of 8.9×10^6 computational cells are employed with mesh refinements near the surface and the trailing edge (Fig. 18). The resolution used for the present results is $\Delta y^+_{min} \approx 2$, $\Delta z^+ \approx 32$, and $\Delta x^+ \approx 87$ at the inlet and $\Delta x^+ \approx 5$ near the trailing edge, see Tab. 3.

Table 3. Parameters and domain of integration for the profile trailing-edge flow simulation. See also Fig. 17

Re_c	$Re_{\delta o}$	δ_o/c	M_∞
$8.1 \cdot 10^5$	15989	0.01972	0.15

L_1	L_2	L_z	grid points
$0.3\ c$	$0.42\ c$	$0.0256\ c$	$8.9 \cdot 10^6$

Δx^+_{min}	Δx^+_{max}	Δy^+_{min}	Δz^+
5	87	2	32

The vortex structures in the boundary layer near the trailing edge and in the near wake are presented by the λ_2 contours in Fig. 19. A complex structure can be observed immediately downstream of the trailing edge. This is due to the interaction of two shear layers shedding from the upper and lower airfoil surface. Note that a small recirculation region occurs right downstream of the trailing edge.

Comparisons of the mean streamwise velocity profiles with experimental data of [19] are presented in Fig. 20 for the upper and lower side, respectively, at several streamwise locations: $x/c = -0.1$, -0.05, -0.02, 0.0 measured

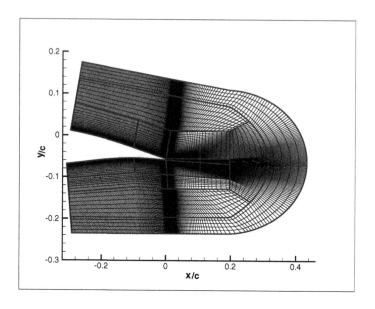

Fig. 18. LES grid of a turbulent flow past an airfoil trailing edge

Fig. 19. Vortex structures in the boundary layer near the trailing edge and in the near wake (λ_2 contours)

from the trailing edge. Whereas in the near wall region a good agreement is observed between the computed and measured mean-velocity profiles. Pronounced deviations occur in the log- and outer region of the boundary layer, which could be caused by the coarsening of the mesh.

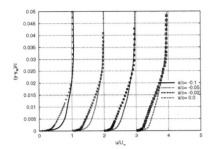
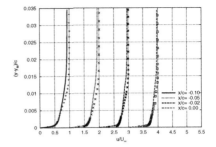

Fig. 20. Mean streamwise velocity profiles near the trailing edge on the upper (left) and lower (right) side compared with experimental data (symbols) [19]

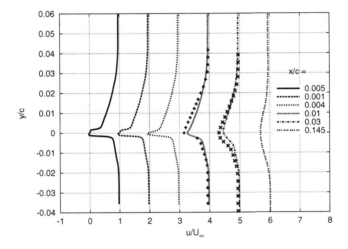

Fig. 21. Mean streamwise velocity profiles in the wake compared with experimental data (symbols) [19]. y/c=0 corresponds to the point where the minimum velocity value in the profile occurs

Further downstream of the trailing edge, asymmetric wake profiles are observed as shown in Fig. 21 in comparison with the experimental data. This asymmetry is generated by varying shear layers on the upper and lower surface shed from the trailing edge. Even at $x/c = 0.145$ no fully symmetric velocity distribution is regained. Note the good qualitative and quantitative experimental and numerical agreement for the velocity distributions.

For the analysis of airfoil flow the skin-friction coefficient is one of the most critical parameters. Its distribution evidences whether or not the flow undergoes separation. Comparisons of the present computations with the experimental values are shown in Figs. 22 on the upper and lower surface, respectively. The simulation results are in good agreement with the data of [19]

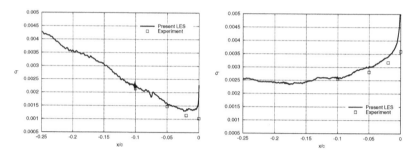

Fig. 22. Skin-friction coefficient on the upper (left) and lower (right) side of the trailing edge compared with experimental data (symbols) [19]

except right at the end of the trailing edge. This could be due to an insufficient numerical resolution near the trailing edge or could be caused by some inaccuracies in the experimental data in this extremely susceptible flow region.

The simulations were carried out on the NEC SX-6 and NEC SX-8 of the High Performance Computing Center Stuttgart (HLRS). The vectorization rate of the flow solver is 99%, and a single processor performance of about 4.3 GFlops on a SX-6 and 6.9 GFlops on a SX-8 processor is achieved. The memory requirement for the current simulation is around 3.5 GB. Approximately 58 CPU hours on 10 SX-8 CPUs for statistically converged solution data are required for the airfoil-flap configuration and roughly 25 CPU hours on 10 SX-8 processors for the zonal approach.

6 Conclusion

An efficient large-eddy simulation method for nearly incompressible flows based on solutions of the governing equations of viscous compressible fluids has been introduced. The method uses an implicit time accurate dual time-stepping scheme in conjunction with low Mach number preconditioning and multigrid acceleration. To validate the scheme, large-eddy simulations of turbulent pipe flow at $Re_\tau = 1280$, and cylinder flow at $Re_D = 3900$ have been performed. The results show the scheme to be efficient and to improve the accuracy at low Mach number flows. Generally, the new method is 6-40 times faster than the basic explicit 5-stage Runge-Kutta scheme.

A large-eddy simulation of the flow field in a tundish is conducted to analyze the flow structure, which determines to a certain extent the steel quality. The findings evidence many intricate flow details that have not been observed before by customary RANS approaches.

The flow over an airfoil with deflected flap at a Reynolds number of Re $= 10^6$ has been studied also based on an LES method. The main characteristics of the flow field are well resolved by the LES. The comparison of the numerical results with the experimental data show a good match of the pressure coefficient distribution and a qualitative agreement of the velocity profiles. The results achieved to date are preliminary but encouraging for further studies. The main reason for the deficiency in the numerical results is the fact, that the resolution requirement for an LES cannot be met everywhere in the computational domain at this high Reynolds number.

The zonal approach results in a pronounced improvement of the local accuracy of the solution. The skin-friction coefficient distribution and the near wall as well as the wake velocity profiles show a convincing agreement with the experimental data. With respect to the simulation of the sound field the LES data from the zonal approach will be postprocessed to determine the source terms of the acoustic perturbation equations, which were already successfully used in [5].

References

1. Turkel E (1999) Annu Rev Fluid Mech 31:385–416
2. Alkishriwi N, Meinke M, Schröder W (2005) Comput Fluids (in press)
3. Gardin P, Brunet M, Domgin JF, Pericleous K (1999) An experimental and numerical CFD study of turbulence in a tundish container. In: Proc. of the 2nd International Conference on CFD in the Minerals and Process Industries CSIRO, Melbourne, Australia
4. Odenthal o, Bölling R, Pfeifer H (2002) Numerical and physical simulation of tundish fluid flow phenomena. In: Proc. of the 11th Japan-German Seminar on Fundamentals of Iron and Steelmaking, Düsseldorf
5. Ewert R, Schröder W (2004) J Sound Vib 270:509–524
6. Wagner S, Bareiß R, Guidati G (1996) Wind turbine noise. Springer, Berlin
7. Howe MS (2000) J Sound Vib 225(2):211–238
8. Meinke M, Schröder W, Krause K, Rister Th (2002) Comput Fluids 31:695–718
9. Rütten F, Meinke M, Schröder W (2001) J Turbulence 2(003)
10. Rütten F, Schröder W, Meinke M (2005) Phys Fluids 17(2)
11. Beaudan P, Moin P (1994) Numerical experiments on the flow past a circular cylinder at sub-critical Reynolds numbers. Technical Report TF-62, Stanford University
12. Davidson L, Cokljat D, Fröhlich J, Leschziner M, Mellen C, Rodi W (2003) LESFOIL: Large eddy simulation of flow around a high lift airfoil. Springer, Berlin
13. El-Askary W (2004) Zonal Large Eddy Simulations of Compressible Wall-Bounded Flows. PhD thesis, Aerodyn. Inst. RWTH Aachen
14. Poinsot TJ, Lele SK (1992) J Comput Phys 101:104–129
15. Ewert R, Meinke M, Schröder W (2002) Computation of trailing edge noise via LES and acoustic perturbation equations. Paper 2002-2467, AIAA
16. Schröder W, Meinke M, El-Askary W (2002) LES of turbulent boundary layers. In: Proc. of the Second International Conf. on Computational Fluid Dynamics ICCFD II, Sydney
17. El-Askary W, Schröder W, Meinke M (2003) LES of compressible wall bounded flows. Paper 2003-3554, AIAA

18. Schröder W, Ewert R (2004) Computational aeroacoustics using the hybrid approach. VKI Lecture Series 2004-05: Advances in Aeroacoustics and Applications

19. Würz W, Guidati S, Herr S (2002) Aerodynamische Messungen im Laminarwindkanal im Rahmen des DFG-Forschungsprojektes SWING+ Testfall 1 und Testfall 2. Inst. für Aerodynamik und Gasdynamik, Universität Stuttgart

Solution of one mixed problem for equation of relaxational filtration by Monte Carlo methods

K. Shakenov

al-Farabi Kazakh National University Al-Farabi av. 71, 050078 Almaty,
Kazakhstan `shakenov2000@mail.ru`

1 Statement of problem

The model of filtration is considered according to the elementary nonequilibrium law in elastic surroundings. The relaxation cores of the filtration law $F(t)$ and the compressibility law $\Phi(t)$ have the following form:

$$F(t) = \frac{\mu}{\kappa}\left\{t + (\tau_w - \tau_p)\left[1 - \exp(\frac{-t}{\tau_p})\right]\right\}\eta(t), \quad \Phi(t) = \rho_0\beta\eta(t),$$

and the model is described by the following system of equations:

$$\chi\Delta\left(p(t,x) + \tau_p\frac{\partial p(t,x)}{\partial t}\right) = \frac{\partial}{\partial t}\left(p + \tau_w\frac{\partial p(t,x)}{\partial t}\right), \tag{1}$$

$$-\frac{\kappa}{\mu}\mathrm{grad}\left(p + \tau_p\frac{\partial p(t,x)}{\partial t}\right) = \mathbf{w}(t,x) + \tau_w\frac{\partial \mathbf{w}(t,x)}{\partial t}, \tag{2}$$

where μ is viscosity of fluid, k is coefficient of penetrability, t is time, τ_w and τ_p are some positive constants of uniformity of time, $\eta(t)$ is Heavisid function,

$$\eta(t) = \begin{cases} 1, & \text{for } t > 0, \\ 1/2, & \text{for } t = 0, \\ 0, & \text{for } t < 0, \end{cases}$$

ρ_0 is density of the fluid in the unperturbed layer conditions, β is elastic capacity coefficient of the layer, $\chi = \frac{\kappa}{\mu\beta}$ is coefficient of piezoconductivity of the layer, Δ is Laplace operator, $p(t,x)$ is pressure, $\mathbf{w}(t,x)$ is filtration velocity vector, [1]. Let us consider the initial boundary value problem for the pressure $p(t,x)$ in a bounded domain $\Omega \in R^3$ with a boundary $\partial\Omega$:

$$\chi\Delta\left(p(t,x) + \tau_p\frac{\partial p}{\partial t}\right) = \frac{\partial}{\partial t}\left(p(t,x) + \tau_w\frac{\partial p(t,x)}{\partial t}\right),$$

$$p(0, x) = 0, \tag{3}$$

$$ap(t, x) + b\frac{\partial p(t, x)}{\partial \mathbf{n}} = \varphi(t, x), \quad 0 < t \leq T, \quad x \in \partial\Omega. \tag{4}$$

where a and b are given constants, $\varphi(t, x)$ is given function. We discretize (1) and (4) only for the temporal variable t, $\quad 0 \leq t \leq T$ with the step $\tau = \dfrac{T}{M}$, $\quad t_n = \tau m$, $\quad m = 0, 1, ..., M$ by implicit scheme. Let us denote $p(t_m, x) \equiv p^m(x)$, $x \in \Omega$ and after the simple transformations we obtain the following finite difference mixed problem for Helmholtz equation with respect to the pressure in the domain $\Omega \in R^3$ for temporal layers $t_n = \tau m$, $\quad m = 0, 1, ..., M$:

$$\Delta p^{m+1}(x) - c p^{m+1}(x) = f\left(p^m(x), p^{m-1}(x), \Delta p^{m-1}(x)\right),$$
$$x \in \Omega, m = 1, 2, \ldots, M - 1, \tag{5}$$

$$a p^m(x) + b\frac{\partial p^m(x)}{\partial \mathbf{n}} = \varphi^m(x), \quad x \in \partial\Omega, \quad m = 0, 1, ..., M, \tag{6}$$

where

$$c = \frac{2\tau_w + \tau}{\chi(\tau_p \tau + 2)}, \quad c > 0, \quad \tau_w > 0, \quad \tau_p > 0, \quad \chi > 0, \quad \tau > 0, \quad R = \chi(\tau_w \tau + 2),$$

$$f\left(p^m(x), p^{m-1}(x), \Delta p^{m-1}(x)\right) = -\frac{4\tau_w}{R} p^m(x) + \frac{2\tau_w - \tau}{R} p^{m-1}(x) +$$
$$\frac{\chi \tau_p \tau}{R} \Delta p^{m-1}(x).$$

2 "Random walk by spheres" algorithm of Monte Carlo methods

The problem (5)–(6) is solved in the same way as the Dirichlet problem for Helmholtz equation, i.e. by the "random walk by spheres" algorithm [2], [3]. But the boundary is considered as a partially absorbing and partially reflecting. When a "particle" hits the ε–boundary $\Big(\varepsilon$ is the vicinity of the boundary $\partial\Omega_\varepsilon$, $\partial\Omega_\varepsilon = \{x \in \Omega; \ d(x) < \varepsilon\}$, $d(x)$ is the distance from the point x to the boundary $\partial\Omega$ of the domain $\Omega \Big)$ of the domain Ω, a "particle" is absorbed with the probability $\dfrac{|\frac{a}{b}|r_i}{1 + |\frac{a}{b}|r_i}$ and reflected to the initial point along the interior normal with the probability $\dfrac{1}{1 + |\frac{a}{b}|r_i}$. Any "particle" reaches the boundary $\partial\Omega$ along the surface normal. Here r_i is the radius of sphere at the moment of reaching the ε – boundary by a "particle", i.e. the distance between the boundary $\partial\Omega$ and the point, which a "particle" reflects in. A "weight" of the boundary, which is proportional to $\left(\dfrac{\varphi(x)}{a}\right)$ and $\left(\dfrac{\varphi(x)}{b}\right)$ respectively, is to be taken into account in absorbing and reflecting [4], [5], [6].

3 "Random walk by lattices" algorithm of Monte Carlo methods

Let e_i be a unit coordinate vector in i–coordinate direction, h is mesh step. Let us denote by H_h the difference mesh on R^3, namely $x \in H_h$, if integer numbers l_1, l_2, l_3 exist, such that $x = \sum_i e_i h l_i$. Let us define $\Omega_h = H_h \cap \Omega$, $\partial\Omega_h =$ {point set on $H_h - \Omega_h$, that situated on the distance of one node from Ω_h along a coordinate direction or a diagonal }. Then, following [7], we approximate the domain Ω and differential operators of (5)–(6) and define the probability of transition at one step of some Markov chain in Ω_h. Let us denote this chain by $\{\xi_n^h\}$ and the transition probability of it by $p^h(x,y)$, which is defined by coefficients of the finite difference equation ([7], page 55, (4.7)–(4.10)).

Further, let us assume, that the chain stops at the first contact with $\partial\Omega_h$. To complete the construction of the chain, which approximates the solution of (5)–(6), it is necessary to discretize the boundary condition (6). For this purpose we assume the following:

$$p_{h(x_i)}(x) = \begin{cases} (p(x + e_i h) - p(x))/h, & \text{if } \tilde{c}_i(x) \geq 0, \quad , x \in \partial\Omega, \\ (p(x) - p(x - e_i h))/h & \text{if } \tilde{c}_i(x) < 0, \quad x \in \partial\Omega, \end{cases} \tag{7}$$

where $\tilde{c}_i(x)$ are the given bounded continued functions ([7], p. 52). Let us define $|\tilde{c}(x)| = \sum_{i=1}^{3} |\tilde{c}_i(x)|$ and the transition probabilities on $\partial\Omega_R^h$:

$$p^h(x, x \pm e_i h) = \frac{\tilde{c}_i^+(x)}{|\tilde{c}(x)|}. \tag{8}$$

It is obvious, that

$$\mathbf{E}_x \left[\xi_{n+1}^h - \xi_n^h | \xi_n^h = y \in \partial\Omega_R^h \right] = \frac{\tilde{c}(y)h}{|\tilde{c}(y)|}. \tag{9}$$

This means, that reflection from the point $y \in \partial\Omega_R$ happens in the direction of $\tilde{c}(y)$. Therefore, we have constructed a Markov chain, that is reflected from $\partial\Omega_R^h$ and stops on $\partial\Omega_A^h$.

Now we return to the total discretization of the equation (5). Let us assume, that $\sup_x \triangle r^h(x) = \sup_x \dfrac{h^2}{Q_h(x)} \to 0$ for $h \to 0$. Then the solution of total finite difference equation $p_h(\cdot)$ is defined by

$$p_h(x) = \exp\left(-c\triangle r^h(x)\right)\left[\mathbf{E}_x p_h(\xi_1^h) + f(x)\triangle r^h(x)\right], \quad x \in \Omega_h. \tag{10}$$

Now we construct the approximation of the solution for $x \in \partial\Omega_R^h$. The approximation (5) gives

$$p_h(x) = \sum_{i,\pm} p^h(x, x\pm e_i h)p_h(x\pm e_i h) + \frac{h\,\varphi(x)}{|\tilde{c}(x)|} + \frac{h\,a}{|\tilde{c}(x)|}p_h(x), \quad x \in \partial\Omega_R^h. \tag{11}$$

Let us denote $d\mu^h(x) = \dfrac{h}{|\tilde{c}(x)|}$, $d\mu_i^h = d\mu^h(\xi_i^h)I_{\partial\Omega_R^h}(\xi_i^h)$, $\mu_n^h = \mu_{n-1}^h + d\mu_{n-1}^h$, $\mu_0^h = 0$.

Let us consider the following approximations:

$$p_h(x) = (1 + ad\mu^h(x))\left[\mathbf{E}_x p_h(\xi_1^h) - \varphi(x)d\mu^h(x)\right], \tag{12}$$

$$p_h(x) = \exp(ad\mu^h(x))\left[\mathbf{E}_x p_h(\xi_1^h) - \varphi(x)d\mu^h(x)\right], \tag{13}$$

$$p_h(x) = \exp(ad\mu^h(x))\mathbf{E}_x p_h(\xi_1^h) - \frac{1 - \exp(ad\mu^h(x))}{-a}\varphi(x), \tag{14}$$

If $d\mu^h(x) \to 0$ for $h \to 0$ uniformly in $x \in \partial\Omega$, then it is possible to use all transition condensations (12)–(14) since this condition is fulfilled, when $\inf |\tilde{c}(x)| > 0$ on $\partial\Omega_R$. If this is not the case, then it is necessary to use (14), as in this case $\triangle r^h(x) \nrightarrow 0$ uniformly in x for $h \to 0$. THe absorbing and reflecting boundaries can be processed simultaneously, if Dirichlet condition $p(x) = a$ is replaced by $a = \dfrac{\varphi(x)}{a}$; $\tilde{c}(x) = 0$ on $\partial\Omega_A$ and (14) is used. But for simplicity, we use (13) and we assume that $\inf\limits_{x\in\partial\Omega_R} |\tilde{c}(x)| > 0$. Let us consider Dirichlet boundary condition:

$$p(x) = a, \quad x \in \partial\Omega_A^h. \tag{15}$$

The equations (10), (13) and (15) form the discretization of (5)–(6). The approximation (13) is chosen for simplification of notations. Now we define

$$A_n^h = \prod_{i=0}^{n}\exp(-c(\xi_i^h)\triangle r_i^h I_{\Omega_h}(\xi_i^h)), \quad C_n^h = \prod_{i=0}^{n}\exp(a(\xi_i^h)d\mu_i^h), \quad D_n^h = A_n^h C_n^h$$

If $N_h = \min\{n : \xi_n^h \in \partial\Omega_A^h\}$, then:

$$p_h(x) = \mathbf{E}_x\left[\sum_{i=0}^{N_h-1} D_i^h f(\xi_i^h)\triangle r_i^h I_{\Omega_h}(\xi_i^h) + D_{N_h-1}^h a(\xi_{N_h}^h) + \sum_{i=0}^{N_h-1} D_i^h \varphi(\xi_i^h)d\mu_i^h\right] \tag{16}$$

is the unique solution of (10), (13) and (15).

In order to demonstrate the possibilities of algorithm we consider a two-dimensional problem in a unit square. The problem is approximated by finite difference method with a constant step h in x and y [7], [8], [9], [10]. Let us show a way of construction of the Markov chain, which approximates the solution, by the algorithm of "random walk by lattices" of Monte Carlo methods. Let (i, j) be the current node. The chain starts from the current node (i, j) and goes from this point into one of neighbouring nodes $(i - 1, j - 1)$, $(i, j - 1)$, $(i + 1, j - 1)$, $(i - 1, j)$, $(i + 1, j)$, $(i - 1, j + 1)$, $(i, j + 1)$, $(i + 1, j + 1)$ with the probability p_l, $(l = 1, 2, ..., 8)$. It is possible to assume that $p_l = 1/8$, $l = 1, 2, ..., 8$, $\sum_{i=1}^{8} p_i = 1$. If the node is on the vicinity of the boundary, and the passage is played into the neighbouring boundary node, then it is necessary to play a new probability of passage into this boundary node. Let $P_b(l)$ be the probability of passage into the boundary node, P_{f-b} be the probability of reflection from the boundary and P_{ab-b} be the probability of absorbing of a "particle" on the boundary. If boundary hit probability has bee played, then after a "particle" reaches the boundary, it will be absorbed with the probability P_{ab-b} and reflected from the boundary with the probability P_{f-b}. If a "particle" has been reflected, then the process of "random walk by lattices" will be continued; the chain breaks, if a "particle" is absorbed on the boundary. If the probability of passage into the boundary node $P_b(l)$ is not played, we continue simulation of the chain from the boundary node. If the probability of reflection from the boundary is played, then a "particle" would be reflected into the previous node. "Weighs" of nodes are calculated along the chain. When the boundary is reached (after absorbing a "particle" on the boundary), the boundary condition is to be taken into account [7], [8], [9]. The theorems of nonbias and finiteness for the constructed estimation can be easily proved. Moreover, the dispersion of the constructed estimate has the explicit expression, [8], [9]. The nonbias of the estimate follows from the fact, that the described "random walk by lattices" algorithm stays within the limits of Neyman-Ulam scheme.

References

1. Molokovich YuM, Osipov PP (1987) Basics of relaxation filtration theory. Kazan University Publishing House, Kazan (in Russian)
2. Elepov BS, Kronberg AA, Mikhailov GA, Sabelfeld KK (1980) Solving of boundary problems by Monte Carlo methods. Nauka, Novosibirsk (in Russian)
3. Ermakov SM, Nekrutkin VV, Sipin AS (1984) Random processes for solving the classical equations of mathematical physics. Nauka, Moskow (in Russian)
4. Haji–Sheikh A, Sparrow EM (1966) SIAM J Appl Math 14(2):370–389
5. Haji–Sheikh A (1965) Application of Monte Carlo methods to thermal coduction problems. Ph.D. Thesis, University of Minnesota, Minneapolis
6. Lindgren B (1962) Statistical Theory. Macmillan, New York

7. Shakenov KK, Musataeva GT (2005) Bulletin of Kazakh National University, series: mathematics, mechanics, informatics 1(44):51–58 (in Russian)
8. Shakenov KK (2002) Comput Technol 7(3):93–97 (in Russian)
9. Ermakov SM, Shakenov KK (1986) Bulletin of Leningrad State University, series: mathematics, mechanics, astronomy: p. 14 ((VINITI) Deposit, No. 6267–B86) (in Russian)
10. Kushner H (1977) Probability methods of approximations in stochastic control and for elliptic equations. Academic Press, New York, San-Francisco, London

Numerical prediction of vortex instabilities in turbomachinery

A. Ruprecht

Institute of Fluid Mechanics and Hydraulic Machinery, University of Stuttgart, Pfaffenwaldring 10, 70550 Stuttgart, Germany ruprecht@ihs.uni-stuttgart.de

1 Introduction

The flow in turbo machinery is mostly characterized by complex geometries and by the presence of a rotating runner or impeller. Additionally it shows a very high Reynolds number, which means that a highly turbulent flow exists. This prevents the possibility to carry out a Direct Numerical Simulation (DNS) or a Large Eddy Simulation (LES). Instead a Reynolds averaged simulation (RANS) is usually applied. For the prediction of unsteady vortices and flow instabilities, however, this approach leads mostly to rather poor results. Recently also a Very Large Eddy Simulation (VLES) approach has been established, which is settled in between of RANS and LES. This approach is discussed in detail in section 3.

Whereas under design conditions the flow in turbo machinery behaves quite smooth, in off-design, however, the flow behaviour is characterized by the existence of strong vortices. These vortices can be unstable and move around in the flow domain. A typical vortex structure in hydro turbines is the draft tube vortex rope. In the design point the velocity distribution downstream of the runner is nearly free of swirl. In part load, however, this distribution shows a high swirling component. This leads to a strong vortex in the draft tube. This vortex can get unstable and form a rotating helical structure, which typically rotates with approximately 30-50% of the runner speed. In figure 1 a typical helical vortex rope is shown in an experiment. The prediction of such an unstable vortex, however, requires a sophisticated modelling, e.g. the standard k-ε model can not predict this phenomenon, see e.g. [1,2]. Applying the standard k-ε model leads to a steady state flow (without a helical vortex). In figure 2 the prediction of the helical vortex is shown, here the adaptive VLES turbulence model described in section 3 is applied.

It can be said, that the prediction of an unsteady vortex motion requires a sophisticated model of turbulence. It also requires an unsteady three-dimensional simulation with a rather fine computational grid. Consequently

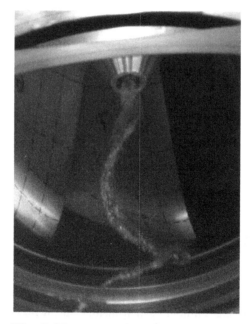

Fig. 1. Vortex rope in experimental set-up

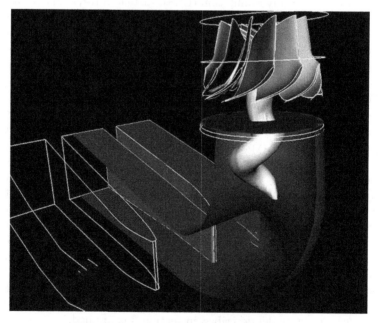

Fig. 2. Numerically predicted vortex rope

the prediction of unstable vortices needs a high computational effort and requires a high computer power. Therefore efficient numerical algorithms, vectorization and parallel computing are required in order to use modern computers efficiently.

The outline of the remaining paper is as follows: In section 2 the numerical methods used are shortly described and its performance on modern computer architectures are discussed. Section 3 contains the description of the VLES approach and the applied turbulence model. In section 4 the numerical procedure and the turbulence model is applied to predict the helical vortex rope in a straight diffuser, which is a simplified draft tube of a hydro turbine. Also the calculation of the unsteady tip vortex of a ship propeller in the wake of the ship is presented.

2 Numerical methods

2.1 Numerical algorithms

The calculations are carried out using the program FENFLOSS which has been developed at the institute for nearly two decades, e.g. [3,4]. The partial differential equations are solved by a Galerkin Finite Element Method. The spatial discretization of the domain is performed by 8-node hexahedral elements. For the velocity components and the turbulence quantities a tri-linear approximation is applied. The pressure is assumed to be constant within each element. For advection dominated flow a Petrov-Galerkin formulation of 2nd order with skewed upwind orientated weighting functions is applied. The time discretization is done by a three-level fully implicit finite difference approximation of 2nd order.

For the solution of the momentum and continuity equation a segregated solution algorithm is used. Each momentum equation is handled independently. The momentum equations are linearized by successive substitution. The linear systems are solved by the BICGSTAB(2) algorithm of van der Vorst [5] with an incomplete LU decomposition (ILU) for preconditioning. The pressure is treated by a modified Uzawa type pressure correction scheme [6]. The pressure correction is carried out in a local iteration loop without reassembling the system matrices until the continuity error is reduced by a given order (usually 6-10 iterations needed).

After the solution of the momentum and continuity equations the turbulence quantities are calculated and a new turbulence viscosity is obtained. The turbulence equations, e.g. k- and ε-equations, are also linearized by successive substitution and the linear systems are solved by the BICGSTAB(2) algorithm with ILU preconditioning. The whole procedure is carried out in a global iteration until convergence is obtained. For unsteady simulations the global iteration has to be carried out in each time step. The iteration procedure is shown in figure 3.

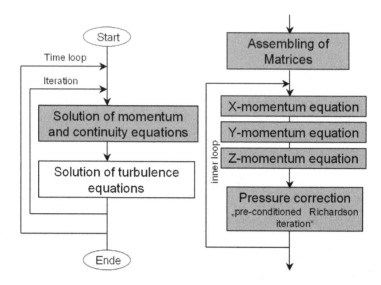

Fig. 3. Iteration scheme of FENFLOSS

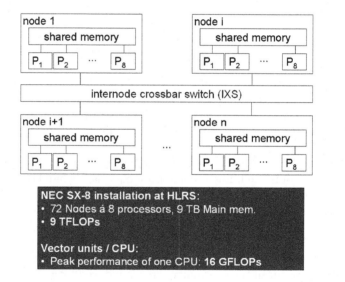

Fig. 4. Architecture and data of NEC SX-8 installation at HLRS

2.2 High performance computing

This type of complex problem requires a high computational effort and can only be handled when applying high performance computing. Most of the calculations are carried out on the NEC SX-8 of the High Performance Computer Center Stuttgart (HLRS). This vector machine has 72 nodes with 8 processors (vector units) each. Within each node there is a shared memory and the nodes are connected by a crossbar switch. The architecture and the performance data are shown in figure 4.

In order to use these kind of machine efficiently FENFLOSS contains distributed memory and shared memory parallelization as well as vectorization. The parallelization of the code is introduced by domain decomposition using overlapping grids. The linear equation solver BICGSTAB(2) is carried out in parallel and the data exchange between the domains is organized on the level of the matrix-vector multiplication in the BICGSTAB(2) solver. The preconditioning is carried out locally on each domain. The data exchange is carried out using MPI (Message Passing Interface) on machines with distributed memory. On shared-memory-computers the code runs also in parallel by applying OpenMP. For details on the numerical procedures and parallelization the reader is referred to [4,7].

Figure 5 shows the speedup of the code. Because of the overlapping grid the maximum speed up is reduced due to the double work to do. The obtained speedup curve of course can not reach this theoretical value because of the overhead and communication. The obtained value shows a very good performance, when each processor handles more than 100000 grid nodes. If the number of nodes per processor is less the overhead and communication get more dominant and the performance drops, see figure 5.

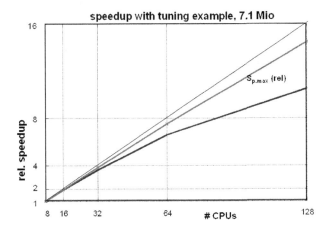

Fig. 5. Speedup of FENFLOSS on NEC SX-8

Each processor consists of several vector units. The peak performance is 16 Gflops per CPU. In FENFLOSS the measured performance reaches 6.4 Gflops, which means 40% of the peak performance. This value is rather good for a Finite Element code using completely unstructured meshes, which means that an indirect addressing for the use of arrays for the sparse matrix operations is necessary.

3 Modelling

3.1 Very Large Eddy Simulation (VLES)

"Real" Large Eddy Simulation (LES) from the turbulence research point of view requires an enormous computational effort since all anisotropic turbulence structures have to be resolved in the computation and only the smallest isotropic scales are modelled. Consequently this method can not be applied for industrial problems today. Today's calculations of flows of practical relevance (characterized by complex geometry and high Reynolds number) are usually based on the Reynolds-averaged Navier-Stokes (RANS) equations. This means that the influence of the complete turbulence behaviour is expressed by means of an appropriate turbulence model. To find a turbulence model, which is able to capture a wide range of complex flow effects quite accurate is impossible. Especially for unsteady flow behaviour this method often leads to rather poor results. The RANS and LES approach can schematically be seen in figure 6, where a typical turbulent spectrum and its division in resolved and modelled parts is shown.

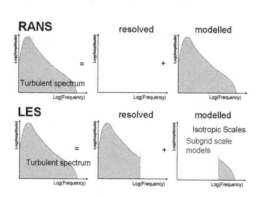

Fig. 6. Modelling approach for RANS and LES

The recently new established approach of Very Large Eddy Simulation can lead to quite promising results, especially for unsteady vortex motion.

Contrary to URANS there is a requirement to the applied turbulence model, that it can distinguish between resolved unsteady motion and not resolved turbulent motion which must be included in the model. It is similar to LES, only that a minor part of the turbulence spectrum is resolved (schematically shown in figure 7). VLES is also found in the literature under different other names:

- Semi-Deterministic Simulation (SDS),
- Coherent Structure Capturing (CSC),
- Detached Eddy Simulation (DES),
- Hybrid RANS/LES,
- Limited Numerical Scales (LNS).

3.2 Adaptive turbulence model

Classical turbulence models, which are usually applied in engineering flow predictions, contain the whole turbulent spectrum. They usually show a too viscous behaviour and very often damp out unsteady motion to early. As discussed above the turbulence model for the VLES has to distinguish between the resolved and unresolved part of the turbulent spectrum (figure 7). Therefore an adaptive model is developed, which adjusts its behaviour according to the approach (schematically shown in figure 8). This means that this model can be applied for all approaches.

The advantage of the adaptive model is, that with increasing computer power the resolved part of the turbulent spectrum increases and the modelled part decreases, consequently the accuracy of the calculations improves.

There are several filtering techniques in the literature [e. g. 8,9]. Here a filtering techniques similar to Willems [8] is applied. In the following the new adaptive turbulence is presented. The nomenclature of resolved and modelled parts can be seen in figure 9.

The basis of this adaptive model is the modified k-ε model of Chen and Kim [10]. This model has been chosen, because it is quite simple and its results

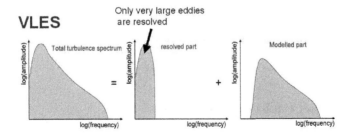

Fig. 7. Turbulent spectrum resolved and modelled by VLES

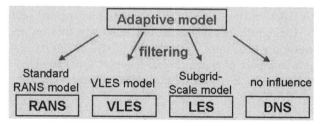

Fig. 8. Adjustment for adaptive model

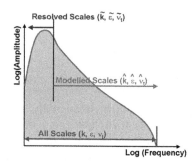

Fig. 9. Distinguishing of turbulence spectrum for VLES

are much better – especially for unsteady flows – compared to the standard k-ε model. The transport equations for k and ε are given as

$$\frac{\partial k}{\partial t} + U_j \frac{\partial k}{\partial x_j} = \frac{\partial}{\partial x_j}\left[\left(\nu + \frac{\nu_t}{\sigma_k}\right)\frac{\partial k}{\partial x_j}\right] + P_k - \varepsilon \tag{1}$$

$$\frac{\partial \varepsilon}{\partial t} + U_j \frac{\partial \varepsilon}{\partial x_j} = \frac{\partial}{\partial x_j}\left[\left(\nu + \frac{\nu_t}{\sigma_k}\right)\frac{\partial \varepsilon}{\partial x_j}\right] + c_{1\varepsilon}\frac{\varepsilon}{k}P_k - c_{2\varepsilon}\frac{\varepsilon^2}{k} + \underbrace{c_{3\varepsilon}\left[\frac{P_k}{k}\right] \cdot P_k}_{\text{additional term}}. \tag{2}$$

These equations prescribe the whole turbulence spectrum and therefore a filtering procedure has to be implemented. According to Kolmogorov theory it can be assumed that the dissipation rate is equal for all scales. This leads to

$$\varepsilon = \hat{\varepsilon} \tag{3}$$

However the turbulent kinetic energy needs a filtering

$$\hat{k} = k \cdot \left[1 - f\left(\frac{\Delta}{L}\right)\right]. \tag{4}$$

A suitable filter can be obtained to

$$f = \begin{cases} 0 & \text{for} \quad \Delta \geq L \\ 1 - \left(\frac{\Delta}{L}\right)^{2/3} & \text{for} \quad L > \Delta \end{cases} \tag{5}$$

where

$$\Delta = \alpha \cdot \max \begin{cases} |u| \cdot \Delta t \\ h_{\max} \end{cases} \quad \text{with} \quad h_{\max} = \begin{cases} \sqrt{\Delta V} & \text{for} \quad \text{2D} \\ \sqrt[3]{\Delta V} & \text{for} \quad \text{3D} \end{cases} \quad (6)$$

with a model constant ($\alpha = 2$ to 5). The volume of the local element is ΔV, u is the local velocity and Δt is the time step. The Kolmogorov scale L is given as

$$L = \frac{k^{3/2}}{\varepsilon}. \quad (7)$$

Introducing the filtering procedure and an appropriate modelling leads to the final equations

$$\frac{\partial k}{\partial t} + U_j \frac{\partial k}{\partial x_j} = \frac{\partial}{\partial x_j} \left[\left(\nu + \frac{\hat{\nu}_t}{\sigma_k} \right) \frac{\partial k}{\partial x_j} \right] + \hat{P}_k - \varepsilon \quad (8)$$

and

$$\frac{\partial \varepsilon}{\partial t} + U_j \frac{\partial \varepsilon}{\partial x_j} = \frac{\partial}{\partial x_j} \left[\left(\nu + \frac{\hat{\nu}_t}{\sigma_k} \right) \frac{\partial \varepsilon}{\partial x_j} \right] + c_{1\varepsilon} \frac{\varepsilon}{k} \hat{P}_k - c_{2\varepsilon} \frac{\varepsilon^2}{k} + \underbrace{c_{3\varepsilon} \left[\frac{\hat{P}_k}{k} \right] \cdot \hat{P}_k}_{\text{additional term}} \quad (9)$$

with the production term

$$\hat{P}_k = \hat{\nu}_t \left(\frac{\partial U_i}{\partial x_j} + \frac{\partial U_j}{\partial x_i} \right) \frac{\partial U_i}{\partial x_j}. \quad (10)$$

For details on the model and its development the reader is referred to Ruprecht [11].

4 Applications

4.1 Swirling flow in a straight diffuser

The VLES approach shown above is applied to swirling flow in a straight diffuser. The results are compared to RANS simulation using the turbulence model of Kim and Chen [9]. Applying the standard k-ε model results in a symmetrical steady state flow, this is in disagreement with experiments and is therefore not considered here. The simulation with the model of Kim and Chen and the VLES approach shows an unsteady flow behaviour. A helical vortex rope appears, which periodically rotates. In figure 10 the flow behaviour is presented. Comparing the different results it can be observed, that using the adaptive model the damping of the vortex is reduced and swirl is less damped out. The vortex is much longer visible. This behaviour corresponds with observation in the experiment, where the vortex exists even longer.

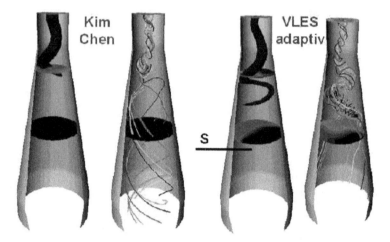

Fig. 10. Predicted vortex rope in a straight diffuser, comparison of RANS (Kim-Chen) and VLES approach

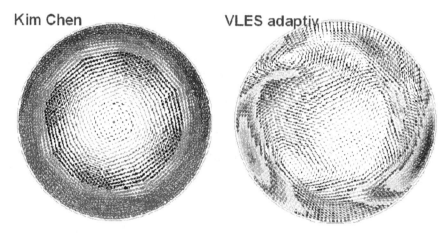

Fig. 11. Predicted secondary velocity in cross-section S, comparison of RANS (Kim-Chen) and VLES approach

Looking at the secondary motion in the cross-sections S (the location can be seen in figure 10), it can be seen, that in the RANS simulation the flow is almost symmetrical, whereas in the VLES large vortex structures can be observed, see figure 11. Again this agrees better with the experiments. Applying the VLES a significant improvement of the calculation accuracy can be obtained.

4.2 Tip vortex at a ship propeller

As an other application the flow around a ship propeller is shown. The flow is unsteady, since the propeller operates in the wake of the ship and consequently does not get a symmetrical inflow. For the simulation only the flow around the propeller is observed, it is calculated in the rotating frame of reference. The inflow boundary condition is taken from velocity measurements of a ship model. This disturbed, steady state velocity field is rotating relative to the propeller. Therefore the flow is unsteady. Special attention is paid to the tip vortex of the propeller. The pressure drop in the vortex can lead to cavitation, which causes noise and vibrations, one of the major problems when designing a new propeller.

In figure 12 the arrangement of the propeller is shown, the ship is equipped with two propellers, in the calculation, however, only one propeller is considered, since their interaction is negligible for the investigation of the tip vortex.

Figure 13 shows the pressure distribution on the blade and an iso-pressure surface at the tip vortex for a certain position of the propeller. It is clearly visible, that in the upper part, where the wake of the ship is more evident, the vortex is stronger than in the lower part. This means, that the strength of the tip vortex at a blade oscillates with the speed of the propeller. In figure 14 the pressure field behind the propeller is shown. It can be observed, that in circumferential direction regions of high pressure, where the blade generates the thrust, are followed by region of low pressure, which correspond to the position of the tip vortices. Again it can be observed, that in the upper part there is a larger pressure drop.

In figure 15 the pressure distribution along the circle shown in figure 14 is plotted. The results of a RANS simulation using the standard k-ε model and the VLES results are compared. The VLES results show a much lower

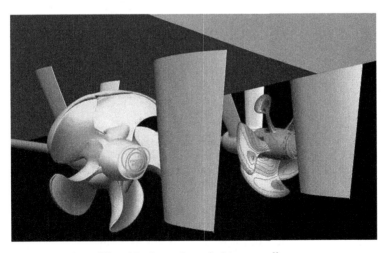

Fig. 12. Investigated ship propeller

Fig. 13. Predicted pressure field and tip vortices with VLES approach

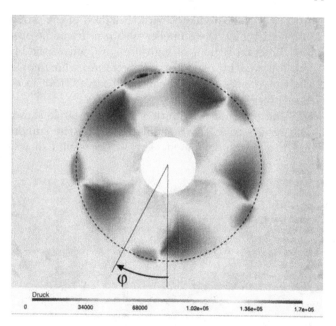

Fig. 14. Pressure field behind the propeller, VLES approach

pressure. At the upper location the pressure goes below the vapour pressure. This means that cavitation would occur (but this is not taken into account in the simulation). In opposition to that the pressure drop obtained by the

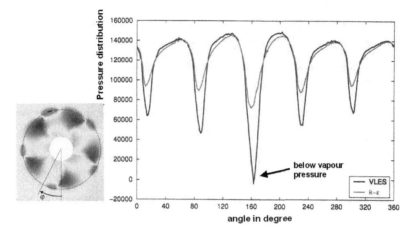

Fig. 15. Predicted Pressure in circumferential direction in radius of tip vortex, comparison of RANS k-ε model and VLES approach

RANS simulation is much lower and does not reach the vapour pressure. This is because the tip vortex is much more damped. From observation in the experiment, however, it is known, that cavitation occurs. This means the VLES results agree much better with these observations.

5 Conclusions

The prediction of unsteady vortex dominated flows show is a challenging task, applying classical RANS simulations often fails and leads to poor results. Here a VLES approach is shown. For this approach an adaptive turbulence model based on the k-ε model of Kim and Chen is developed. Applying this approach to vortex dominated flows - vortex rope in a straight diffuser and tip vortex at a ship propeller - leads to an increase on computational accuracy. Therefore this approach seems to be suitable for this types of flows.

Acknowledgement: The author wants to thank his colleagues at the Institute for the cooperation, especially Felix Lippold, Thomas Helmrich and Martin Maihöfer, which partly carried out the computations.

References

1. Ruprecht A, Helmrich T, Aschenbrenner T, Scherer T (2002) Simulation of vortex rope in a turbine draft tube. In: Avellan F, Ciocan G, Kvicinsky S (eds) Proc. of the 21st IAHR Symposium on Hydraulic Machinery and Systems, Lausanne

2. Scherer T, Faigle P, Aschenbrenner T (2002) Experimental analysis and numerical calculation of the rotating vortex rope in a draft tube operating at part load. In: Avellan F, Ciocan G, Kvicinsky S (eds) Proc. of the 21st IAHR Symposium on Hydraulic Machinery and Systems, Lausanne

3. Ruprecht A (1989) Finite Elemente zur Berechnung dreidimensionaler turbulenter Strömungen in komplexen Geometrien. Ph.D. thesis, University of Stuttgart, Stuttgart

4. Maihöfer M (2002) Effiziente Verfahren zur Berechnung dreidimensionaler Strömungen mit nichtpassenden Gittern. Ph.D. thesis, University of Stuttgart, Stuttgart

5. Van der Vorst HA, (1994) Recent developments in hybrid CG methods. In: Gentzsch W, Harms U (eds) Proc. High Performance Computing and Networking, München

6. Zienkiewicz OC, Vilotte JP, Toyoshima S, Nakazawa S (1985) Comp Meth Appl Mech Eng, 51:3–29

7. Maihöfer M, Ruprecht A (2003) A local grid refinement algorithm on modern high-performance computers. In: Chetverushkin B, Ecer A, Periaux J, Satofuka N, Fox P (eds) Proc. of Parallel CFD 2003, Elsevier, Amsterdam

8. Willems W (1997) Numerische Simulation turbulenter Scherströmungen mit einem Zwei-Skalen Turbulenzmodell. Ph.D. thesis, Shaker Verlag, Aachen

9. Magnato F, Gabi M (2000) A new adaptive turbulence model for unsteady flow fields in rotating machinery. In: Proc. of the 8th International Symposium on Transport Phenomena and Dynamics of Rotating Machinery (ISROMAC 8)

10. Chen YS, Kim SW (1987) Computation of turbulent flows using an extended k-ε turbulence closure model, NASA CR-179204

11. Ruprecht A (2005) Numerische Strömungssimulation am Beispiel hydraulischer Strömungsmaschinen. Habilitation thesis, University of Stuttgart, Stuttgart

Notes on Numerical and Fluid Mechanics and Multidisciplinary Design